Contents

DESIGNING A TECHNOLOGY PROJECT

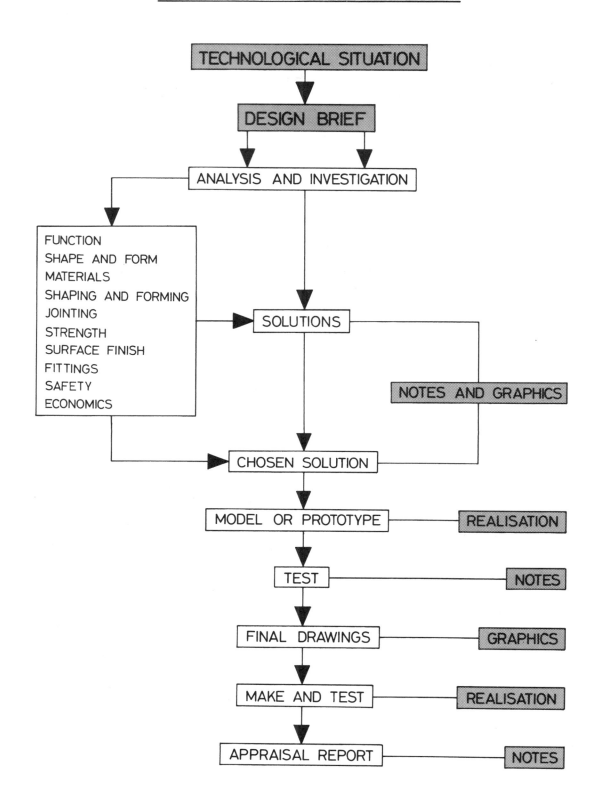

TECHNOLOGICAL SITUATION

DESIGN BRIEF

ANALYSIS AND INVESTIGATION

FUNCTION
SHAPE AND FORM
MATERIALS
SHAPING AND FORMING
JOINTING
STRENGTH
SURFACE FINISH
FITTINGS
SAFETY
ECONOMICS

SOLUTIONS

NOTES AND GRAPHICS

CHOSEN SOLUTION

MODEL OR PROTOTYPE — REALISATION

TEST — NOTES

FINAL DRAWINGS — GRAPHICS

MAKE AND TEST — REALISATION

APPRAISAL REPORT — NOTES

Design and Technology

A. Yarwood
A. H. Orme

HODDER AND STOUGHTON

LONDON   SYDNEY   AUCKLAND   TORONTO

Preface

My book 'Design and Craft', published in 1979, was written in collaboration with S. Dunn. The book has been well received in schools and this present book, 'Design and Technology' shows how the methods advocated in the previous book can be applied to the teaching of school technology.

The two authors of this book have both had considerable experience in developing and teaching school technology courses in comprehensive schools. A. Yarwood, while a member of the Crafts, Applied Science and Technology (CAST) subjects committee of the Schools Council during the 1970s, developed a course in technology at the High Ridge Comprehensive School, Scunthorpe, where he was then the Head of Department of Technical Drawing and Technology. A. H. Orme is at present the Head of Graphics and Technology at the Thomas Alleyne's High School, Uttoxeter. The methods described in this book have been developed and fully tested in his department there. Both have also had considerable examining experience. A. H. Orme has examined in Technical Drawing and in crafts, as well as acting as a visiting moderator for crafts with a CSE examinations board. He is also a member of the Technical Subjects panel of the board and a member of a committee responsible for introducing Control Technology as an Ordinary level examination with a GCE examinations board. A. Yarwood is a Chief Examiner for Graphical Communication with a University School Examinations Department at Ordinary level, has, until this year, been a Chief Examiner for Technical Drawing with a CSE examinations board as well as being a visiting moderator/examiner in Design and Craftwork for another GCE examining board.

One aim of this book is to show how a design process can be applied to the solution of projects in school technology. A second aim is to demonstrate that a school technology course can be fully developed from design and craft courses. The actual making of the 'hardware' associated with a technological problem at school can, indeed does, involve reasonable practical skills. A problem which can confront many teachers who wish to introduce technology courses into their school curriculum, is the extraordinary width that the term 'technology' can appear to cover. In this book technology has been closely associated with design and craft. Technology is concerned with the applications of scientific knowledge to the practical solutions of problems. Technology is therefore basically a design or problem solving process. It is the practical applications of scientific knowledge and the problem solving methods of a design process with which this book is concerned.

A. Yarwood.

British Library Cataloguing in Publication Data

Yarwood, A.
   Design and technology,
   1. Design—Study and teaching (Secondary)—England
   2. Technology—Study and teaching—England
   I. Title   II. Orme, A. H.
   607'.42   NK1170

   ISBN 0 340 32975 0

First printed 1983
Fifth impression. 1987

Typeset in 10/11 pt. Univers (Monophoto) by Macmillan India Ltd., Bangalore 25

Printed in Hong Kong for
Hodder and Stoughton Educational,
a division of Hodder and Stoughton Ltd.,
Mill Road, Dunton Green, Sevenoaks, Kent TN13 2YD,
by Colorcraft Ltd

Cover

The cover photograph is of computer graphics designed and produced by The Moving Picture Company.

SECTION 1

Introduction

Designing a technology project

The flow chart on page 4 shows the stages in designing a technology project. Details concerning some of the stages in the **design process** shown by the flow chart will be given later in this book. Shaded statements indicating the activities associated with the design process have been added to the flow chart. The essential activities associated with designing are—making notes, making graphical representations of the design and realising (actually making) the design.

Technological situation
Any design, whether technological or otherwise, is the result of attempts at finding solutions to a problem. The whole process of designing is based on the need to solve problems. Take as an example the bedside light project described in pages 14 to 17. This project involves the practical application of scientific knowledge—of electricity and electrical circuits—to a given situation—the need for some form of extra lighting in a bedroom.

Design brief
In any given situation there must be several ways of arriving at a solution to the problems involved. The design brief states quite clearly how the situation is to be resolved. The design brief should take the form of a written statement describing what is to be designed. Again from the example of the bedroom light project—the design brief for this states quite clearly that the design is to be in *unit* form, with its *own* power source and that it is to provide *temporary* lighting.

Analysis and investigation
Before solutions to a design brief can be attempted, problems associated with designing the project will require analysing and investigating. As some of the problems are solved, solutions may be attempted. As the investigation proceeds further solutions may be seen and attempted. Ten areas requiring investigation are given in the flow chart. There may well be others when designing for some projects. The areas of investigation may be taken in any suitable order and some may need much more attention than others. Notes, and sometimes tests, will usually form part of the investigation.

Solutions
While considering the various aspects of the investigation, a number of solutions will arise. Notes and drawings should indicate the flow of ideas for solving the design brief. The solutions will normally take the form of drawings and notes describing how the design is developing.

Chosen solution
From the ideas shown by the variety of solutions, a single solution will be chosen. This should satisfy the demands of the situation and the design brief.

Model or prototype
When a solution has been chosen, a model, or scale prototype—even a full size prototype—should be considered. Such models may show up design problems not appreciated or foreseen when writing notes and making drawings.

Test
If possible, the model should be tested to find out whether the design functions well.

Final drawings
Final drawings made with the chosen solution drawings, notes and tests as a guide, should always be produced before making the design. At this drawing stage further difficulties may be seen and it is easier to correct a drawing than it is to correct a completed project.

Make and test
Now make the design.

Appraisal report
Does the design meet the design brief requirements? Check, test and report on the design. Modifications to improve the design can be suggested. Do not be content to state that you find the design completely satisfactory. The vast majority of finished projects can always be improved.

Examples of high technology

Concorde taking off on a passenger flight. Photograph by courtesy of British Aerospace.

The Humber Bridge. Photograph by courtesy of Freeman Fox and Partners, Consulting Engineers.

British Rail's Advanced Passenger Train. Built mainly of light alloys to save weight, the train has an aerodynamically shaped nose to reduce wind resistance. Photograph by courtesy of the British Rail Board.

The radiotelescope at Parkes, New South Wales, Australia. 63 metres in diameter. Photograph by courtesy of Freeman Fox and Partners, Consulting Engineers.

Dr. Stewart Reddaway with the pilot model of the Distributed Array Processor (DAP) which he originated at ICL's Research and Advanced Development Centre at Stevenage, England. Photograph by courtesy of International Computers Limited (ICL).

Technology in industry

The six photographs on this page were taken at the works of J. C. Bamford Excavators Limited of Rocester, Staffordshire.

The authors would like to place on record here their appreciation to J. C. Bamford Limited for allowing them to take and reproduce these six photographs.

1. A design problem to be solved

4. Testing parts of the design

2. A model of the design

5. Preparing a working drawing

3. Making a full size prototype

6. A completed machine ready for export

Examples of school technology

Ten examples of school technology projects are described in pages 10 to 44. These commence with two projects suitable for pupils early on in their fourth year in a secondary school and progress to a final and comparatively complex project suitable for a group of fifth year pupils. Each project starts with a statement of a situation from which a design brief is chosen. This is followed by an analysis and investigation in note form. These investigations take the form that is given in the design process chart on page 4. Drawings and notes then describe each project. These show possible solutions and a chosen solution to the design brief. A large variety of graphical techniques are illustrated in the ten examples, combined with different methods of presentation. Photographs accompanying the descriptions of the projects show either the completed design or stages in reaching each completed design.

The intention of showing such a variety of graphical methods is to indicate to readers how design folders or folios may be produced. The reader can choose which methods he or she prefers to adopt. Because of the limitations of space, some of the descriptions in pages 10 to 44 may be incomplete. The intention in these pages is to show how projects can be 'written up' in notes and drawings, rather than to show fully completed project folders.

The ten projects

Project 1 Centres of gravity
Pages 10 and 11. An example of a project based on an information sheet which has been given to pupils by a teacher. The information sheet briefly describes what is meant by a centre of gravity (C of G) and goes on to explain methods of determining the centre of gravity of a lamina. The simple project derived from the information sheet is one which could not function properly unless the centre of gravity of the 'CRDBTOIL' is found. The sheet of drawings showing solutions for an outline shape for a 'CRDBTOIL' (Counter Rotating Dynamically Balanced Tension Operated Irregular Lamina) has been drawn in pencil. Crayon shading or colour paint washes have been added to accentuate the shapes of some of the drawings. Notes have been included with the drawings using a black ball-point pen.

Project 2 Structures
Pages 12 and 13. An example of a project devised by a teacher to introduce pupils to the concept of structures. The problem posed is to construct a strong structure using a minimum of materials. The idea of testing is also introduced in this project. Two sheets of drawings, both drawn with ink pens on A3 size paper, show possible solutions together with a final solution. The

drawings on the first sheet have been drawn freehand, those on the second sheet have been drawn with the aid of a straightedge. Similarly the notes on the first sheet are in freehand lettering, while on the second sheet a lettering stencil has been used. Both sets of drawings show leg ties shaded in with a pencil. In the final solution, these leg ties were found to be unnecessary, the structure standing up well to its imposed test.

Project 3 Bedroom light
Pages 14 to 17. A project involving a simple electrical circuit. A variety of techniques are involved in the presentation of the graphics associated with this project. The left hand of the two columns of drawings on page 15 show electrical circuit diagrams drawn with an ink technical pen (a 'Rotring' pen) and with the aid of drawing instruments. The right hand column shows outline pictorial sketches drawn freehand with a black ball-point pen. The drawings on page 16 show freehand sketches in pencil on an A2 size sheet. Various shading techniques employing coloured pencil crayons and colour washes are included to emphasise the pencil sketches. Finally the working drawing, page 17, involves instrument drawing, freehand pictorial drawing, a 'ghosted' drawing and the use of figure and letter stencils. This project lays particular emphasis on the fact that a large number and variety of solutions are possible when considering even the simplest of technological projects.

Project 4 100 km/h vehicle
Pages 18 to 21. This project offers a challenge to the student. Incidentally, one which is impossible of achievement. In attempting to achieve the objective of the project however—to reach 100 km/h—much interesting information can be gathered. Solutions to the problems arising involved velocity, acceleration, movement of fuel in a tank, propeller speeds, frame construction, among other problems requiring solutions. The graphics show instrument drawings with pencil shading, freehand pencil drawings with colour wash shading and a pencil working drawing together with notes added to drawings with stencils, freehand in pencil and freehand using a black 'penstik' pen.

Project 5 Automatic door-opening device
Pages 22 to 25. A project involving a pneumatic system, but one which could have been solved using a variety of other methods. The drawings on page 23 are freehand, outline, pen drawings involving no shading. Eight suggestions for solving the door opening problem are shown, together with details of the chosen method—of using a pneumatics system. When making up this solution, it was discovered that an extension piece was needed on the cylinder ram to allow the system to function. Pen drawings made with the aid of instruments are shown on page 24. These have, in part, had colour wash shading added. The circuit drawings on page 25 show two different symbol types. CETOP

symbols are now being adopted by British Standards. However many circuit drawings will still be seen using KAY symbols.

Project 6 Walking-on-water
Pages 26 to 29. A project which could have been solved using a variety of methods, some of which are shown in the outline pen drawings on pages 27 and 28. These were drawn on sheets of A4 unlined paper, each being numbered for filing in a folder. The working drawing of the chosen solution was drawn on an A3 sheet with a technical pen using instruments and stencils. A shaded, pencil drawing on page 29 shows details of the construction of one 'shoe' of the device. All shading was added with water colour paints applied by brush.

Project 7 Dramatics effects box
Pages 30 to 33. Another project offering a wide range of solutions. The chosen solution involved what was probably the most simple answer—to operate the various effects with pull strings taken to backstage. Suggestions for solutions are shown in a series of numbered A4 sheets, some freehand, some drawn with the aid of instruments. Notes have been added to these drawings, either freehand or with stencilled letters. Colour wash shading has been adopted where necessary to emphasise the drawings. The drawings numbered 8 and 9 on page 33 show the use of coloured lines to emphasise the working parts of mechanisms in pictorial drawings. These two are instrument drawings with lettering drawn with the aid of stencils.

Project 8 Three-minute timer
Pages 34 to 36. A project solved by using an electronics circuit and involving an integrated circuit chip. In this example, although a number of possible solutions are offered, it became obvious almost from the start that an electronics system offered the best solution. Note the method, shown on page 34, of showing graphically how one idea follows another. All the graphics associated with this project have been drawn in ink using a technical pen, instruments and letter stencils. Coloured lines and shading have been employed to bring to notice those details requiring emphasis.

Project 9 Hovercraft
Pages 37 to 39. An example of a project in which the solution was already known—the principles involved in the well-known hovercraft. The problems involved were therefore only those of elements of shape, size, position and angles. The graphics are all on A4 sheets, except the working drawing of page 39. This is on an A3 sheet. All drawings were produced in ink with 'penstik' Indian ink pens in two colours. Notes have been written with a black ball-point pen. The cover of the A4 folder in which the graphics are kept has a title and name printed with 'Letraset' letters. 'Letraset' is a dry-transfer process of lettering.

Project 10 Aero-generator
Pages 40 to 44. An ambitious project undertaken by a group of pupils at the Walton High School, Stafford. Although 5 pages have been used to describe this project, some details concerning the experiments and research carried out to solve the problems involved, have not been included. However sufficient has been included to show the general lines of the research and how the problems involved were eventually solved. A variety of graphical methods have been used in producing drawings describing the project—freehand and instrument drawing, freehand and stencil lettering, colour shading, colour line emphasis, use of square grids for circuit drawings, pen and pencil drawing.

Note
Most of the work shown in the ten projects is the result of projects undertaken in schools where school technology is practiced. Some of the drawings will therefore either be incomplete or perhaps contain small errors. The intention of including these ten projects in this book has been to show how solutions to technological problems can be achieved and to show how the designing process can be applied to school technology. It is inevitable that exact and precise answers cannot be found using the methods practised in schools. The descriptions of the projects do however show how interesting, varied and even exciting, the designing of school technology projects can become.

Project folders or folios
Throughout this book it is suggested that all notes and graphics associated with a project be kept in a folder or folio. An example of a simple folder cover is shown below.

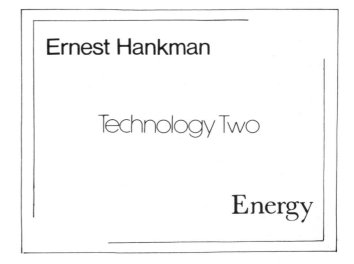

Project 1 Centres of gravity

Situation
You have been given a study sheet entitled 'Centres of Gravity'.

Design brief—Design a CRDBTOIL—Counter Rotating Dynamically Balanced Tension Operated Irregular Lamina.

Analysis and investigation
Function—A toy with a rotating and counter rotating movement.

Shape and form—Many different shapes are suitable. However some shapes may be unsafe, tending to break up when rotating. Thus some shapes, e.g. a ship with masts, are unsuitable.

Materials—Any craft sheet material capable of withstanding the stresses involved e.g. sheet plastics, aluminium, plywoods, of a thickness not exceeding 3 mm.

Shaping and forming—Various. Curve cutting saws, various smoothing techniques, casting, laminating, bending.

Jointing—Except for laminating, preferable not to use joints from the safety aspects of the joint breaking as the device rotates.

Strength—Must be able to withstand high peripheral speeds and constant changes of direction.

Surface finish—Smooth—high speeds and safety.

Fittings—Strong, nylon based cord.

Safety—Note—if cord breaks nothing happens, but if device breaks, flying sheet pieces may be a source of danger.

Economics—Very low cost.

Special factors—Cord holes 5 mm from centre of gravity. Holes must only just be clearance diameter of cord used. Investigate 'old' toys. Colour can vary with rotation.

Anything, whether natural or man-made (e.g. a spanner), may be thought of as being made from a vast number of tiny but equal particles. Each particle is pulled towards earth by GRAVITY.	By experiment it is possible to BALANCE the spanner on the tip of a finger. There is only one point at which this can happen. This point is called the CENTRE OF GRAVITY.	What has happened to make the spanner balance is that at the centre of gravity the tendency to turn in one direction is counteracted by the tendency to turn in the other. In more exact terms the spanner is in EQUILIBRIUM.	In fact TWO FORCES are acting on the spanner in equilibrium on your finger-tip. The force of gravity acting vertically downwards, and the EQUAL and OPPOSITE reaction of your finger acting vertically upwards.

DEFINITION: The centre of gravity of a body is defined as the point of application of the resultant force due to the earth's attraction on it.

Methods of finding the centre of gravity of a lamina
Your finger-tip is really quite large ($\emptyset$6 mm) and if you now tried the balancing experiment again but this time using a needle point ($\emptyset$0.001 mm) you would find it very difficult indeed to find the precise centre of gravity.

1
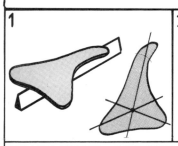

2

Exercises
1. Find out what a RESULTANT is.
2. Explain what happens when your finger-tip is supporting the spanner but is not under the centre of gravity. Explain why.
3. Find out where the centre of gravity is in a square, rectangle, circle and in a triangle.

STUDY SHEET—CENTRES OF GRAVITY.	Amanda Sanbrooke Group 2 27 Feb 87

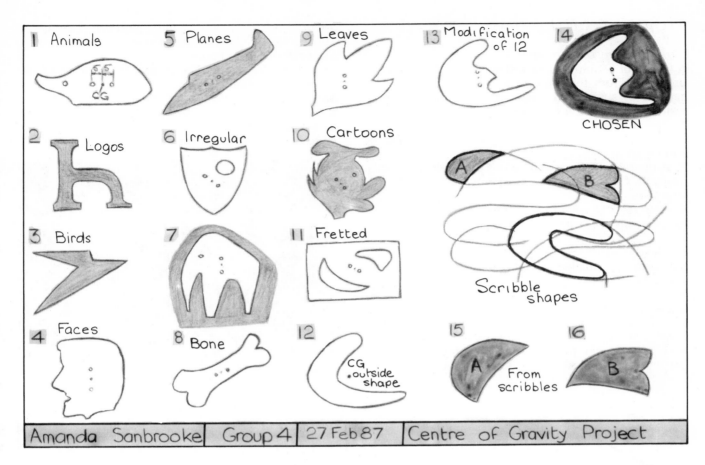

1 Animals
5 Planes
9 Leaves
13 Modification of 12
14 CHOSEN

2 Logos
6 Irregular
10 Cartoons

3 Birds
7
11 Fretted
Scribble shapes

4 Faces
8 Bone
12 CG outside shape
15 A From scribbles
16 B

| Amanda Sanbrooke | Group 4 | 27 Feb 87 | Centre of Gravity Project |

Average peripheral speed of a CRDBTOIL

The movement of a CRDBTOIL is one of constant acceleration and retardation. Calculation of the maximum peripheral speed is therefore not easy. However it is possible to calculate the average peripheral speed.

Method

1. Rotate the CRDBTOIL and count the number of pulls taken in 10 seconds.
2. Collapse the device in its wound state and count the number of turns in the loop.
3. Measure the distance from the C of G of the device to its furthest edge. Then:

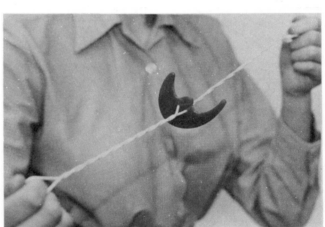

Number of pulls in 10 s = 22;
Number of turns = 15;
Centre of gravity to edge = 48 mm.
Pulls in 1 minute = 22 × 6 = 132
Revolutions per minute = 132 × 15 × 2 = 3960

$$\text{Peripheral speed} = \frac{\text{rev/min} \times \pi D \times 60}{1\,000\,000} \text{ km/h}$$

$$= \frac{3960 \times 3.14 \times 48 \times 60}{1\,000\,000}$$

$$= 35.81 \text{ km/h}$$

This calculation was based on a cord loop of 500 mm. Using various lengths of loop calculate how they affect the peripheral speed.

11

Project 2 Structures

Situation and design brief

A length of 50 mm square softwood, 300 mm long, is provided. From this make a structure 250 mm high, capable of supporting a seated adult. The following restrictions are to be observed:

1. The structure is to be freestanding.
2. The only other materials allowed will be pva glue and masking tape.
3. A platform 300 mm square of 18 mm blockboard is available for testing. This must not form part of the structure.
4. Your teacher will cut the softwood into strips on a machine saw as requested.

Analysis and investigation

Function—A challenge—to produce an efficient structure.
Shape and form—Any—strength of more importance.
Materials—Stated in brief. Use the minimum possible.
Shaping and forming—When the strips required are sawn from the 50 mm square section, it must be remembered that there will be considerable waste because of the thickness of the saw cuts.

Jointing—Joints must be purpose designed. Traditional joints may weaken the structure.
Strength—Of paramount importance.
Surface finish—None. The strips can be left as sawn.
Fittings—None allowed.
Safety—Large strong top for testing. Must not test by sitting until bricks or weights have been employed as a preliminary test.
Economics—Not really applicable.

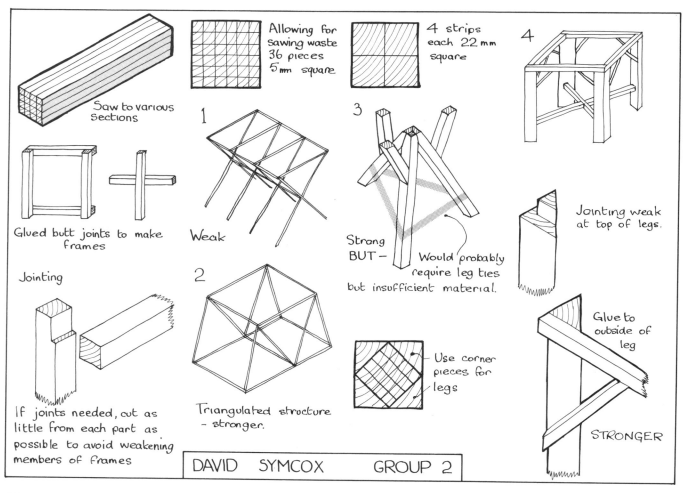

Saw to various sections

Allowing for sawing waste 36 pieces 5 mm square

4 strips each 22 mm square

4

Glued butt joints to make frames

Jointing

1 Weak

2 Triangulated structure – stronger.

3 Strong BUT – Would probably require leg ties but insufficient material.

Use corner pieces for legs

Jointing weak at top of legs.

Glue to outside of leg

STRONGER

If joints needed, cut as little from each part as possible to avoid weakening members of frames

DAVID SYMCOX   GROUP 2

Appraisal

The structure was made exactly as designed. When tested it was thought that the structure was rather 'stiff'. This 'stiffness' could possibly be reduced by slight amendments to the positions of the crossbraces. The structure was tested by a teacher whose weight was equivalent to 520 newtons downwards when sitting. As he moved slightly on the structure, it was obvious that, except for the legs which were always in compression, the other parts were sometimes in tension and sometimes in compression.

Weight/strength ratio

Weight of original 300 mm by 50 mm square = 3 N

Weight of strips sawn from original wood = 2.1 N

Completed structure (shaded parts unused) = 1.2 N

$$\text{Proportion of wood lost in sawing} = \frac{3 - 2.1}{3} = \frac{0.9}{3}$$

$$= 30\%$$

$$\text{Proportion of wood used in structure} = \frac{1.2}{3} = 40\%$$

Total weight of structure + platform (8 N) = 9.2 N

$$\text{Weight/strength ratio} = \frac{9.2}{520} = 1:57$$

$$\text{Ignoring platform—Weight/strength ratio} = \frac{1.2}{520}$$

$$= 1:433$$

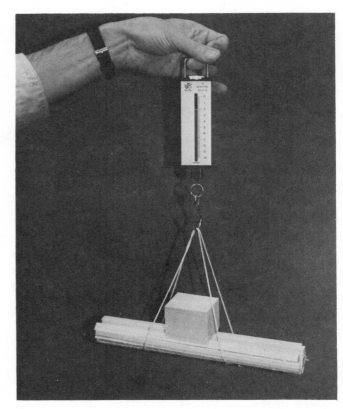

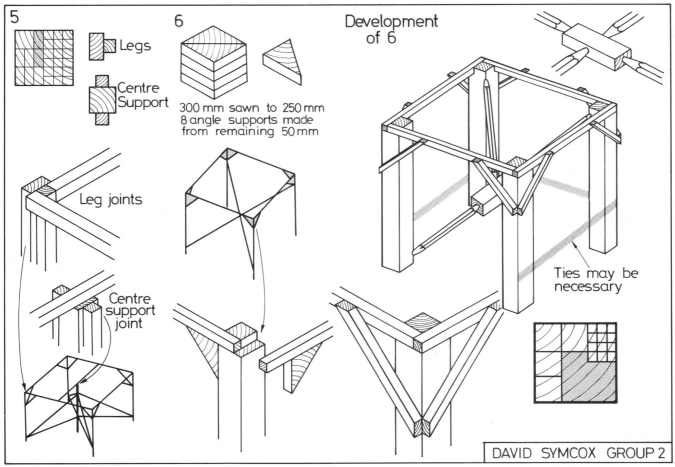

5

Legs

Centre Support

6

300 mm sawn to 250 mm
8 angle supports made
from remaining 50 mm

Leg joints

Centre support joint

Development of 6

Ties may be necessary

DAVID SYMCOX GROUP 2

Project 3 Bedroom light

Situation
A bedroom has only one light switch, located near the door. There is no power socket in the room. It is necessary to get into bed in the dark, after switching the light off, or get out of bed in the dark to switch the light on.

Brief
Design a light unit, with its own power source, suitable for temporary lighting in the bedroom.

Analysis and investigation
Function—To give 'general' lighting while:
(a) moving across the bedroom;
(b) looking at the clock etc. during the night.
Shape and form—See drawings on page 16.
Economics—Make a costs list of all parts with prices taken from up-to-date catalogues and stating the costs of fittings and electrical parts. 'Long life' batteries, although more expensive when purchased, are more economical in use.
Shaping and forming—Wooden case jointed. Acrylic sheet cover head bent after edge jointing opaque and textured parts. Make cover first and shape sides of case to the curves of the cover.
Jointing—Housing joints for box. The three parts of the acrylic sheet cover edge jointed with acrylic cement. Metal parts of circuit screwed to casing.
Materials and fittings

No.	Item	Material or fitting
1	Case	10 mm hardwood
2	Cover	3 mm opaque acrylic sheet
3	Cover	3 mm clear textured acrylic sheet
4	Reflector	Tinplate
5	Contacts	Brass strip
6	Switch	Latching press switch
7	Screws	10 mm × 4 Round head
8	Switch spring	Polyurethane foam
9	Battery	1.5 V
10	Bulb	1.25 V 0.25 A
11	Feet	10 mm square rubber 3 mm thick
12	Cable	0.2 mm insulated copper wire

Strength—The unit will be switched on and off in the dark. Unit must thus be sufficiently strong to stand up to rough treatment which may result from this.
Surface finish—Cover through which light shines to be a textured finish—scatters light and allows fingers location in dark.
Special factors—Unit must be easy to locate and switch in the dark.
Safety—All edges must be rounded. Unit must be stable on bedside table/cabinet. No earthing necessary—voltage too low.

Electrical circuits

Suggestions for basic circuits
(a) A circuit with switches independent of the unit.
(b) A self-contained unit with included switch.
(c) A combination of (a) and (b).
Circuit 1—This circuit causes two problems:
(a) Switch B must be closed to operate light from the door.
(b) Long cables would be required between the unit and the switch near the door.
Circuit 2—This circuit allows either the switch near the door or the switch on the unit to operate the unit light ON or OFF. This circuit causes two problems:
(a) Double pole switches are more expensive than single pole switches.
(b) Long cables would be required between the unit and the door.
Circuit 3—Simple circuit with a single switch on unit. A toggle switch as shown in this circuit may however cause the user of the unit problems in locating the switch in the dark.
Circuit 4—A press switch in place of the toggle switch of circuit 3. *Note*—there are two types of press switch—a *momentary* switch and a *latching* switch. A latching switch allows the light to remain ON when the switch is pressed, and is switched OFF when the switch is pressed a second time.
Circuit 5—A more complicated circuit. With switch A ON the full brightness of the light is obtained. With switch B ON a dimmed light is obtained.
Note—Circuit 4 was used in the final solution.

Drawings
A variety of methods of presentation are shown in the drawings associated with the notes for this project. Circuit diagrams on page 15 have been drawn with the aid of instruments. Ergonomics drawings on page 15 have been drawn freehand with a black ball point pen. Drawings on page 16 have been drawn freehand with pencils. Shading has been included with coloured crayons and water colour paints. The working drawing on page 17 has been drawn in ink with the aid of instruments and lettering stencils. Part of the working drawing has been made freehand in ink.

Appraisal
The finished design is easy to use as a free-standing unit and the light it gives is adequate, even though it shines upwards. The design is attractive and works well.

The acrylic cover must be made before the case. This enables the wooden sides of the case to be shaped to the curves of the cover—an easier operation than attempting to shape the case sides to the cover.

The large, easy-to-operate press switch may have other applications e.g. as an emergency switch, as an alarm switch.

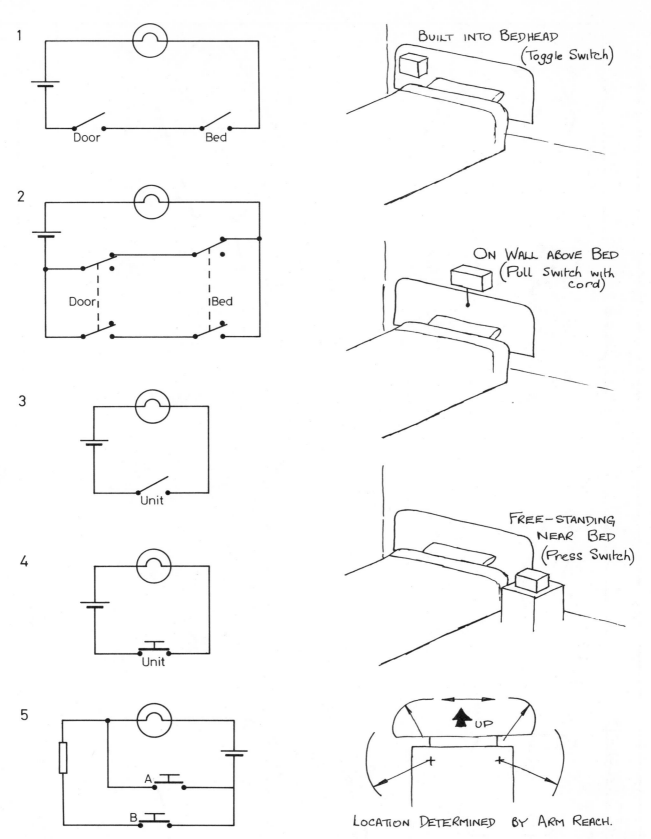

1

Door   Bed

BUILT INTO BEDHEAD
(Toggle Switch)

2

Door   Bed

ON WALL ABOVE BED
(Pull Switch with
cord)

3

Unit

FREE-STANDING
NEAR BED
(Press Switch)

4

Unit

5

A

B

UP

LOCATION DETERMINED BY ARM REACH.

Bedroom light circuit diagrams (drawn with instruments)

Bedroom light ergonomics drawings (drawn freehand)

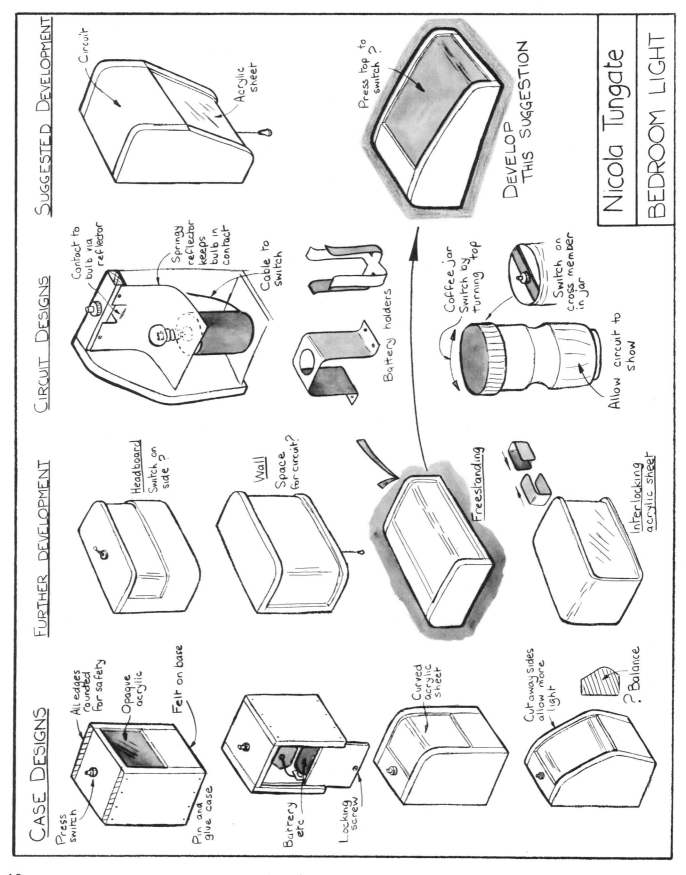

SUGGESTED DEVELOPMENT

Circuit

Acrylic sheet

Press top to switch ?

DEVELOP THIS SUGGESTION

Nicola Tungate

BEDROOM LIGHT

CIRCUIT DESIGNS

Contact to bulb via reflector

Springy reflector keeps bulb in contact

Cable to switch

Battery holders

Coffee jar
Switch on by turning top

Switch on cross member in jar

Allow circuit to show

FURTHER DEVELOPMENT

Headboard
Switch on side ?

Wall
Space for circuit?

Freestanding

Interlocking acrylic sheet

CASE DESIGNS

All edges rounded for safety

Opaque acrylic

Felt on base

Press switch

Pin and glue case

Battery etc

Locking screw

Curved acrylic sheet

Cutaway sides allow more light

? Balance

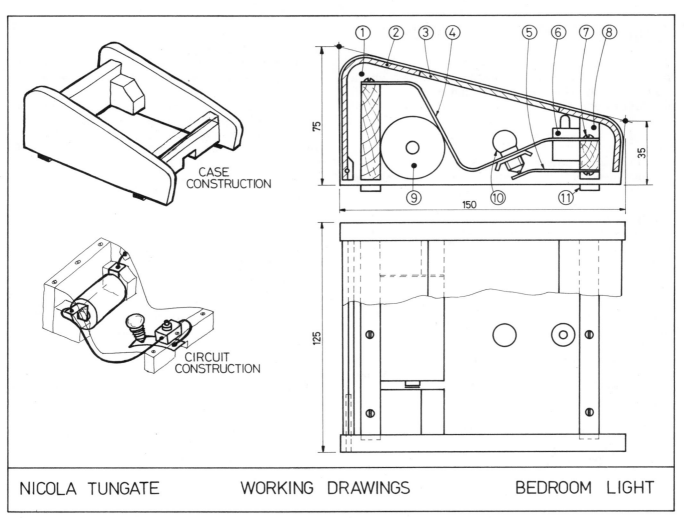

CASE
CONSTRUCTION

CIRCUIT
CONSTRUCTION

75

35

150

125

① ② ③ ④ ⑤ ⑥ ⑦ ⑧

⑨ ⑩ ⑪

NICOLA TUNGATE   WORKING DRAWINGS   BEDROOM LIGHT

Project 4   100 km/h vehicle

Situation
A model diesel engine is available. It is fitted with a 200 × 150 propellor. The engine can produce 11 000 revolutions per minute (rev/min) when fixed to a test bed.

Brief
Design a vehicle powered by the diesel engine to attain a target speed of 100 kilometres per hour (km/h). The vehicle must retain contact with the ground at all times.

Analysis and investigation
Function—The only function of the design is to attempt attaining the maximum speed of 100 km/h under the power of the diesel engine.

Shape and form—This is a purely utilitarian design and only functional design aspects need to be considered. The appearance of the resulting vehicle need *not* be considered. The shape and form of the vehicle will be governed by:
(a) the centre of gravity (C of G) of the vehicle;
(b) the weight of the engine and its vibration when working—the vehicle must carry this weight and be capable of resisting the vibration;
(c) air resistance when the vehicle attains a high speed;
(d) the dangers inherent when the vehicle moves at high speed.

Materials and fittings

Size	Material	Item
800 mm	Ø3 mm mild steel rod	Frame
200 mm	Ø3 mm silver steel rod	Wheels shaft
2	Ø50 mm 'pneumatic' wheels	Wheels
4	Washers	Wheels
4	Wheel collars	Wheels
100 mm × 70 mm × 10 mm	Hardwood	Aerofoil
4	M6 bolts and nuts	Engine mounting
	Tinplate	Fuel tank
	Ø3 mm pipe (plastic)	Fuel lead
	Ø3 mm pipe (copper)	Fuel tank
6 metres	7 strand heavy duty control line wire	Tether
600 mm	Ø25 mm mild steel rod	Tethering pylon
2	M12 nuts	,, ,,
2	M12 washers	,, ,,
1	Ball bearing	,, ,,
1	Oil drum	,, ,,
	Sand and cement	,, ,,

Economics—Make a costs list of all parts shown in the materials and fittings table. All costs should be taken from up-to-date catalogues. Remember to add the costs of those materials and fittings used in experimental vehicles which are discarded.

Shaping and forming—Frame made from 3 mm mild steel rod, each part being bent to shape as required. Wheel shaft made from 3 mm silver steel rod.

Jointing—Wire frame parts are brazed to each other. Engine is bolted onto the engine mounting platform. The petrol tank is soft soldered.

Strength—Essential for safety. If in doubt, sacrifice performance in order that the vehicle is sufficiently strong.

Special factors—This is an experimental project. The final design will be produced only after several 'trial and error' models have been tested.

Safety—With a propellor mounted on an engine capable of revolving at 11 000 rev/min, the running of the vehicle can be dangerous. *Every* precaution in production and trials must be taken. Double check all parts of the design before running the vehicle.

Engine and propellor
On a test bed the engine can attain 11 000 rev/min.

The propellor specification is 200 × 150. The 200 part of this specification means that the tips of the propellor blades revolve in a circle of 200 mm diameter. The 150 part of the specification is the *pitch* of the propellor. When the propellor makes one revolution in still air, it can *theoretically* move a vehicle 150 mm.

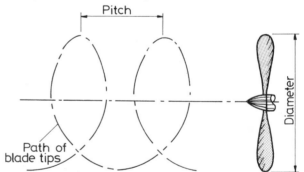

From the engine speed and the propellor specification it is possible to calculate the *theoretical* speed which can be attained by a vehicle fitted with the engine and propellor.

$$\frac{\text{Engine rev/min} \times \text{propellor pitch} \times 60 \text{ (minutes)}}{1000 \text{ (metres)} \times 1000 \text{ (kilometres)}}$$
$$= \text{km/h}$$

$$= \frac{11\,000 \times 150 \times 60}{1000 \times 1000}$$

$$= \frac{11 \times 15 \times 6}{10} \quad \text{(after cancelling out)}$$

$$= 99 \text{ km/h} \ (\textit{theoretical maximum speed})$$

However, with propellor 'slip' and the inevitable air resistance of the vehicle, it is doubtful whether this maximum speed can be attained.

Tethering of vehicle

For safety reasons it is considered that the vehicle must be tethered to a central post, about which the vehicle will run in a circular path.

Note on circular path

For safety reasons, the vehicle is to be tested for speed along a circular path. As a result, the aim of achieving 100 km/h will be impossible. This is because any body travelling along a circular path, even at what appears to be a constant speed, is in fact accelerating. This acceleration is known as *centripetal acceleration*.

Exercise

With the aid of text books on theoretical mechanics explain:

(a) why is energy used to cause a body to accelerate;
(b) what is meant by centripetal acceleration;
(c) why does centripetal acceleration prevent the achievement of the aim of 100 km/h.

Let the required circular path be 25 metres long.

$$\text{Circumference} = 2 \times \text{Radius} \times \pi$$
$$\therefore \ 25 = 2 \times R \times 3.14$$
$$\therefore \ R = \frac{25}{2 \times 3.14}$$

∴ Tether cable = 3.98 metres long.

Running on this path of 25 metres circumference

$$1 \text{ revolution per second} = \frac{25 \times 60 \times 60}{1000} \text{ km/h}$$

$$= 90 \text{ km/h}$$

∴ 1 revolution per second = 90 km/h
2 seconds per revolution = 45 km/h
3 seconds per revolution = 30 km/h
and so on.

Also 100 km/h = 1 circuit in 0.9 s
or 10 circuits in 9 s.

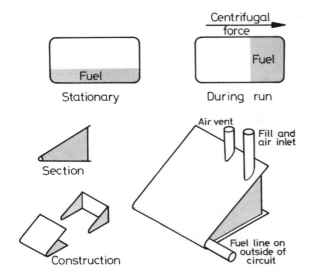

Stationary   Centrifugal force   During run

Section   Construction

Air vent
Fill and air inlet
Fuel line on outside of circuit

Tethering cable

7 strand wire cable as used for heavy control lines for model aircraft.

Tethering points on vehicle

First models to have three tether points. Cables from the three tether points must be adjusted in length to a common join point.

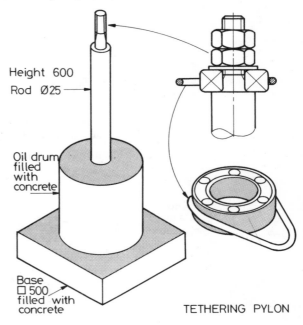

Height 600
Rod Ø25
Oil drum filled with concrete
Base □ 500 filled with concrete

TETHERING PYLON

Wheels

Three types of wheel were tested—all of 50 mm external diameter.
'Streamline'—hard rubber tyres with metal alloy hubs. The tyres roll off at speed.
Treaded 'pneumatic'—rubber tyre with alloy and plastics hubs. Gave best performance and were used in the vehicles tested.
Plastics tyred—heavy metal hubs with plastics tyres. Gave good results, but transmit much surface vibration from circuit to vehicle.

"Streamline"   "Pneumatic"   Plastics tyred

Fuel tank

1. When the vehicle runs at speed on its circular path, centrifugal force will move all the fuel in the fuel tank to the outer wall of the tank.
2. Tank design must allow for this movement. Otherwise the engine will cut-out because fuel will not be able to run to the engine.
3. The tank design is therefore based on a wedge shape.

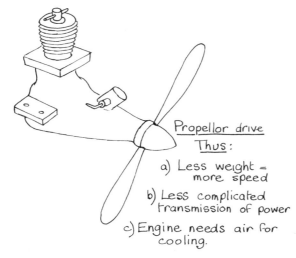

Propellor drive
Thus:
a) Less weight = more speed
b) Less complicated transmission of power
c) Engine needs air for cooling.

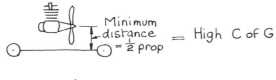

Minimum distance = $\frac{1}{2}$ prop = High C of G

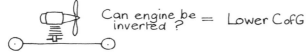

Can engine be inverted? = Lower C of G

First ideas

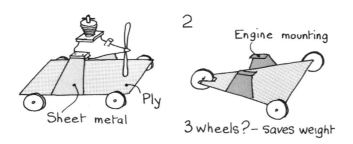

Sheet metal   Ply

2
Engine mounting
3 wheels? - saves weight

3
Engine forward?
Use Meccano strips?

4
Wire frame?

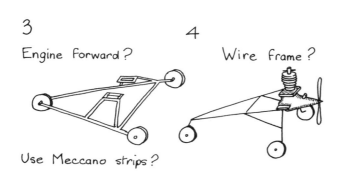

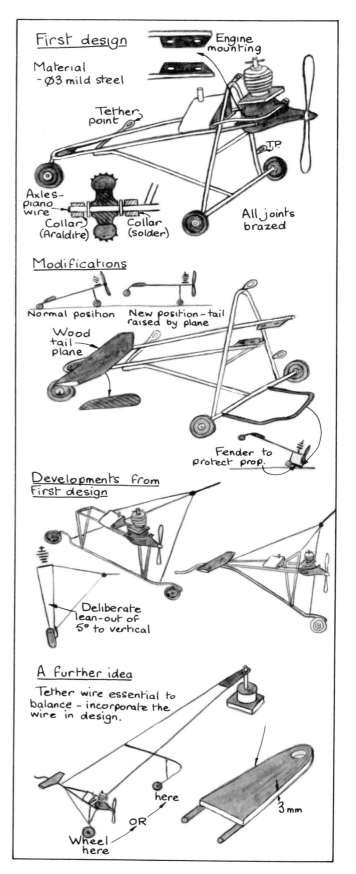

First design
Material - Ø3 mild steel
Engine mounting
Tether point
Axles-piano wire
Collar (Araldite)   Collar (solder)
All joints brazed
T.P.

Modifications
Normal position   New position-tail raised by plane
Wood tail plane
Fender to protect prop.

Developments from First design
Deliberate lean-out of 5° to vertical

A further idea
Tether wire essential to balance - incorporate the wire in design.
here
OR
Wheel here
3mm

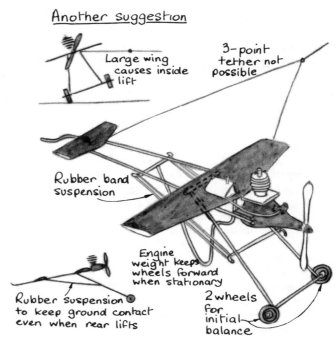

Large wing causes inside lift

3-point tether not possible

Rubber band suspension

Engine weight keeps wheels forward when stationary

Rubber suspension to keep ground contact even when rear lifts

2 wheels for initial balance

Note on drawings

A variety of drawing techniques are shown in the drawings associated with this project. The drawings on page 19 have been drawn with the aid of instruments, templates and stencils. Drawings on page 20 have been drawn freehand, either in ink or in pencil, with crayon and water colour shading. The upper drawings on this page continue the same technique as on page 20. The lower drawing on this page is a working drawing made in ink.

Note on this project

The description of this project includes only the work carried out on the 'Mark 1' suggestions. This design was not fully satisfactory and a maximum speed of only 39.5 km/h was obtained. Further development of the project would have produced vehicles capable of attaining better speeds.

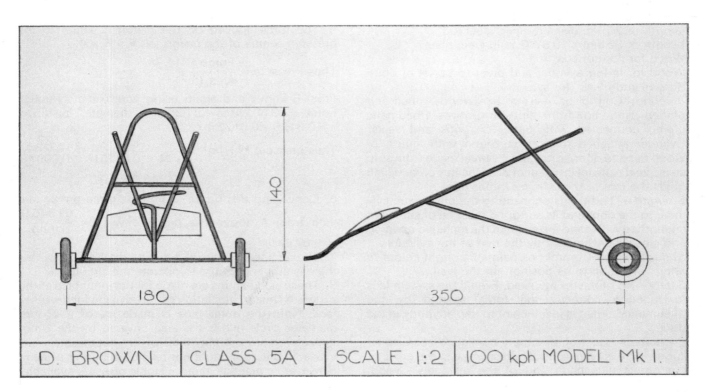

| D. BROWN | CLASS 5A | SCALE 1:2 | 100 kph MODEL Mk I. |

Project 5 Door opening device

Technological situation
In a technology workshop the storeroom is in constant use by pupils. The door of the storeroom cannot be easily opened by one person when equipment is being carried in and out of the store. 13 ampere electric sockets and an air line are available near the store room.

Design brief
Design a switch operated device for opening and closing the door. When opened the door must remain open until the operator decides to close it.

Analysis and investigation
Function—The device must operate the door in the sequence: operate switch—door opens—stays open—operate switch—door closes.
Shape and form—Providing the design functions well and the door opens fully to 90°, the shape and form do not matter to any degree. However, the design should be unobtrusive and should not depend upon alterations to the storeroom or surrounding walls. In the chosen solution the storeroom door had to be modified in order to allow the device to open the door through 90°.
Materials and fittings—Double acting pneumatic cylinder—diameter 25 mm; stroke 60 mm.
Adaptor to fit plastic piping to existing air line.
2 × 3 port valves.
1 × 5 port valves.
1 × flow restrictor.
Lengths of plastic piping, with nuts, 'O' rings and seals.
2 Tee joints for piping.
Lengths of 40 mm × 3 mm mild steel strip.
Lengths of 20 mm × 3 mm mild steel strip.
Length of 20 mm diameter mild steel rod.
Length of 50 mm × 16 SWG mild steel plate.
Wood for control box.
Wood for fitting cylinder and pivot to portal of door.
Screws and rivets (for swivels).
Jointing—Control box—8 mm hardwood—glued and pinned joints; box front slides in grooves. Pneumatic piping connected with nuts, 'O' rings and seals. Cylinder attached to door surround with length of wood fixed to door portal and a swivel made from strip steel; steel plate let into rear of door locates pivot which holds the ram of the cylinder in position.
Strength—The fittings connecting the pneumatic cylinder to the door and its surround must be of sufficient strength to withstand the strain of the repeated opening and closing of the door by the ram of the cylinder.
Surface finish—Control box painted a bright colour to attract attention to its position on the wall.
Safety—No problems involved. Even if the system fails to function correctly the force exerted by the pneumatics circuit is insufficient to trap anybody in the door.
Economics—Cost of air supply negligible because air supply pipe is always under pressure. List the costs of the pneumatics parts and of the materials—wood, metal, screws etc. and estimate their cost from up-to-date catalogues.

Notes on the pressures required to operate the system
1. By experimenting with a spring balance, it was found that 200 grams force (2 N) was required to open the door when applied at the handle—700 mm from hinges.
2. The stroke of the ram of the pneumatic cylinder employed is 60 mm. If the door is to open through 90°, the ram must pivot 42 mm from the hinges. By Pythagoras applied to a right angled triangle of 60 mm hypotenuse: square on hypotenuse = 60^2 = 3600; square on each other side = $3600 \div 2$ = 1800. $\therefore$ other sides each $\sqrt{1800}$ = 42.45 giving the approximate length of 42 mm.
3. A force of 2 N at 700 mm produces a moment of 1400 Nm about the hinges. At 42 mm from the hinges, a force of $1400 \div 42$ = 33.3 N is required to produce the necessary turning moment.
4. The present British Standards unit of pressure for pneumatic circuits is the lbf/in² (pounds force per square inch). It may be necessary to convert pounds force per square inch units to SI units. The SI unit of pressure is the pascal (Pa). 1 Pa = 1 newton/square metre or N/m².
Taking *approximate* figures:
1 kg = 2.2 lb and 1 kgf = 10 N.
Then 1 lb = 0.45 kg and 1 lbf = 4.5 N.
Also 1 m = 39.4 in. Thus 1 m² = 1552.36 in².
Thus, if a pressure of 1 lbf/in² is applied to each of the square inches over a square metre, the pressure in pascals over the square metre = 1552.36 × 4.5 N m² = 6985.62 = approximately 7000 N m².
5. The force applied on the piston is equal to the pressure × area of the piston, or, $F = P \times A$.

$$\text{Thus Pressure} = \frac{\text{Force (N)}}{\text{Area (m}^2\text{)}}$$

From 3 above and again using approximate figures:
Force = 33 N; Area of 25 mm diameter piston = $\pi \times 0.0125 \times 0.0125$ m².

$$\text{Thus pressure in N/m}^2 = \frac{33}{3.14 \times (0.0125)^2} = \frac{33}{0.000\,49}$$

$$= 67\,346.9 \text{ N/m}^2.$$

6. Converting this figure to pounds force per square inch from 4 above the pressure will be $\dfrac{67\,346.9}{7000}$ = 9.62 lbf/in².
7. The air line operates at a minimum of 40 lbf/in², so there is ample pressure to operate the circuit.
8. These calculations are made for the inner face of the piston. A similar calculation could be made for the outer face. Note the outer face is made up of a 25 mm diameter circle minus the area covered by the 8 mm diameter ramrod. If only 9.62 lbf/in² of pressure were available the door could not be closed by this system. However, in practice, there is ample pressure available.

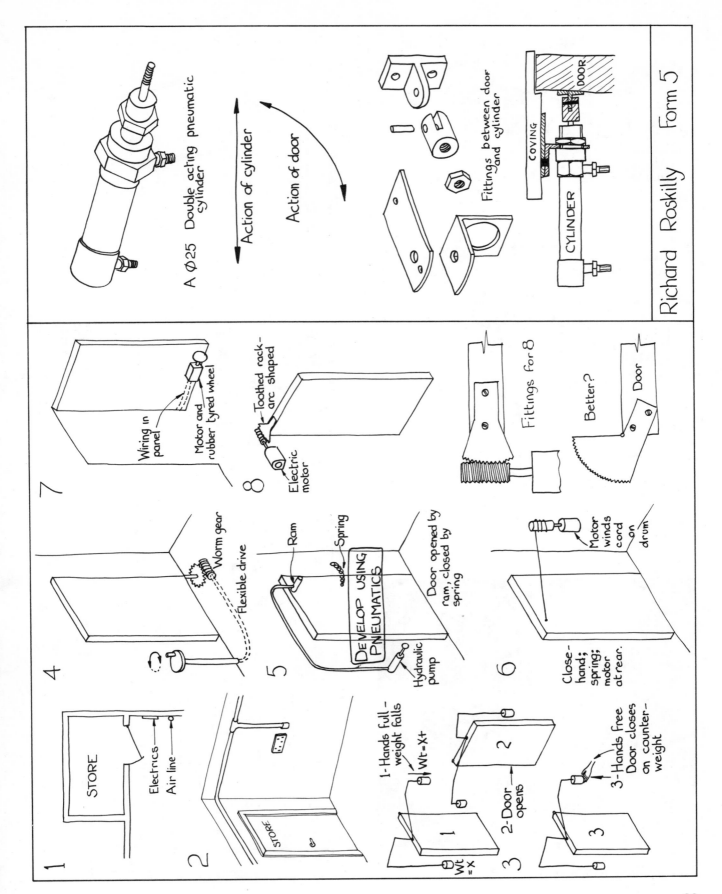

A Ø25 Double acting pneumatic cylinder

Action of cylinder

Action of door

Fittings between door and cylinder

COVING

DOOR

CYLINDER

Richard Roskilly   Form 5

1

STORE

Electrics

Air line

2

STORE

3

1- Hands full - weight falls

Wt = X+

Wt = x

1

2

2- Door opens

3

3- Hands free Door closes on counter-weight

4

Worm gear

Flexible drive

5

Ram

Spring

DEVELOP USING PNEUMATICS

Hydraulic pump

Door opened by ram, closed by spring

6

Motor winds cord on drum

Close- hand; spring; motor at rear.

7

Wiring in panel

Motor and rubber tyred wheel

8

Toothed rack- arc shaped

Electric motor

Fittings for 8

Better?

Door

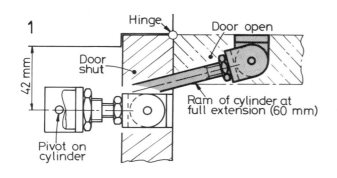

1 Hinge — Door open — Door shut — 42 mm — Pivot on cylinder — Ram of cylinder at full extension (60 mm)

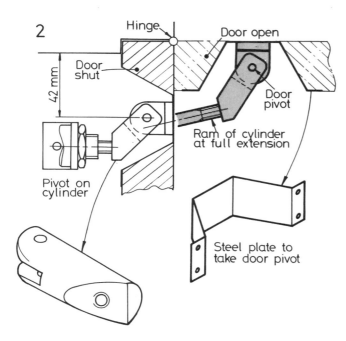

2 Hinge — Door open — Door shut — 42 mm — Door pivot — Ram of cylinder at full extension — Pivot on cylinder — Steel plate to take door pivot

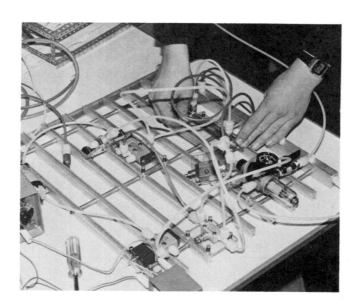

Experimenting with a circuit

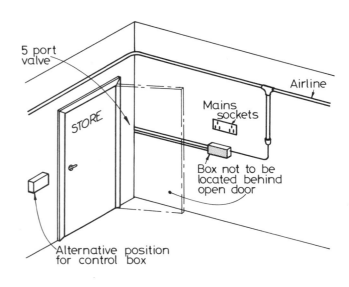

5 port valve — STORE — Mains sockets — Airline — Box not to be located behind open door — Alternative position for control box

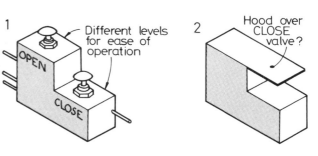

1 OPEN — CLOSE — Different levels for ease of operation

2 Hood over CLOSE valve?

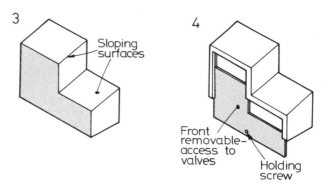

3 Sloping surfaces

4 Front removable-access to valves — Holding screw

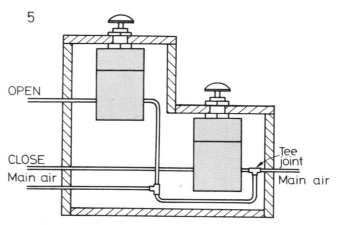

5 OPEN — CLOSE — Main air — Tee joint — Main air

24

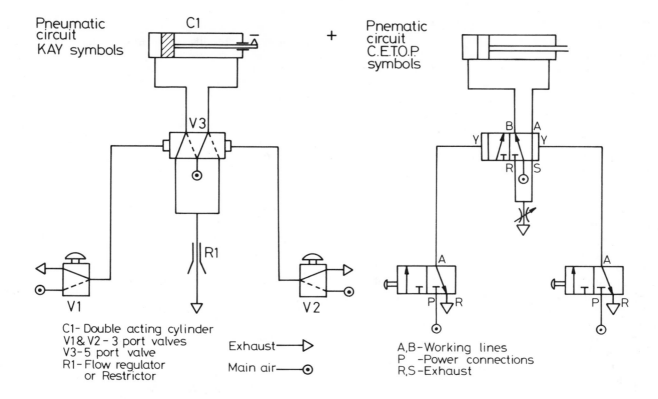

Pneumatic circuit KAY symbols

Pnematic circuit C.E.T.O.P. symbols

C1 - Double acting cylinder
V1 & V2 - 3 port valves
V3 - 5 port valve
R1 - Flow regulator
    or Restrictor

Exhaust ▷
Main air ⊙

A,B - Working lines
P - Power connections
R,S - Exhaust

9. *Note* all calculations involve approximations *and* frictional losses have been ignored.

Circuit operation

1. Valve V1 pressed—allows air pressure to operate Valve V3.

2. V3 changes—allows air pressure to operate the cylinder C1 moving piston *out*.

3. As piston moves out air in C1 behind the piston escapes through V3 but is restricted by restrictor R1. This restrictor of air escape controls speed of piston movement and thus the speed of the door opening movement.

4. Circuit is complete—door is open. If V1 is pressed again, nothing happens.

5. Valve V2 pressed—causes C1 piston to move *in*. Movement *in* again controlled by R1 restrictor. Door closes at same speed as it opened.

6. Press V2 again—nothing happens.

A double acting pneumatic cylinder is shown on the right

Project 6 Walking-on-water

Situation
A small cabin boat cannot be moored alongside the bank of a river. Regulations at the mooring site allow no boats to be moored nearer than 50 metres from the bank, except to pick up passengers and stores.

Design brief
Design a device for walking on water which will enable a person to get to the boat from the river bank.

Analysis and investigation
Function—The device must easily support the mass of a human body. It must be capable of being carried to the mooring site by a person wishing to reach the boat.
Shape and form—The shape of the design must be such as to allow easy movement through water. The mass of water displaced by the design must be at least as great as the mass of a person using it—preferably greater to allow a margin of safety.
Materials—The chosen design is to be made from 2 mm WPB plywood with 8 mm WPB plywood supports and 8 mm thick softwood strips.
Shaping, forming and jointing—A lightweight frame to be made from 8 mm plywood and 8 mm thick softwood slats to which a 2 mm skin of plywood is glued. Edge joints of the skin to be sewn with copper wire and sealed with fibre glass mat and waterproof polyester resin.
Surface finish—Surface grain to be filled in, rubbed down and then several coats of paint to be applied. A waterproof paint must be chosen for the purpose.
Fittings—Pivots to be made to hinge the paddle. These must be made from brass and fitted with brass screws.
Safety—Two safety aspects had to be considered in the chosen design. First, *each* float must be capable of sustaining the weight of the person using them. Second, the floats must be stable. If unstable then user may be tilted either forwards or sideways into the water.
Economics—A complete list of all the items involved in making the design should be drawn up and costs taken from up-to-date catalogues.

Problems which need to be overcome
1. Sideways stability can be assisted by designing and making a pair of 'water sticks'.
2. The 'sliding' action of walking on water means that it is very difficult to attain forward momentum.
3. There exists the possibility of uncontrollable sideways sliding to a 'splits' position of the legs. Some control over this can be gained by practice in using the floats and by use of 'water sticks'.
4. A large float is not really necessary. The density of water is 1 gram per cubic centimetre and the law of flotation states: A floating body displaces its own weight of the fluid in which it floats. Thus a man of average weight of 65 kg is of a weight 65 × 1000 grams = 65 000 grams. To support an average weight man would therefore require a float of volume 65 000 cubic centimetres (cm^3). By calculation using a graphical method (not shown here), it was found that the area of the deck of the float is approximately 7560 cm^2. The average depth of the float is approximately 17 cm. Thus the volume of the float is 7560 cm^2 × 17 cm = 128 520 cm^3, nearly twice that required to support a man of average weight.

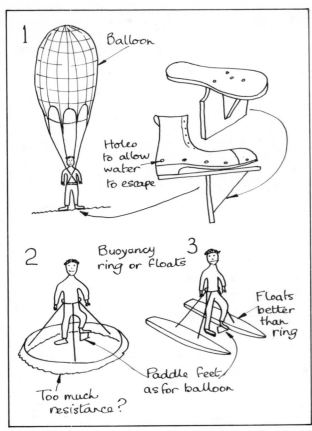

1 Balloon
Holes to allow water to escape

2 Buoyancy ring or floats
Too much resistance?

3 Floats better than ring
Paddle feet, as for balloon

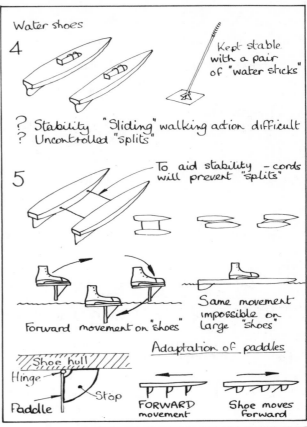

Water shoes

4 Kept stable with a pair of "water sticks"
? Stability   "Sliding" walking action difficult
? Uncontrolled "splits"

5 To aid stability – cords will prevent "splits"
Forward movement on "shoes"
Same movement impossible on large "shoes"

Adaptation of paddles
Shoe hull
Hinge
Paddle   Stop
FORWARD movement   Shoe moves forward

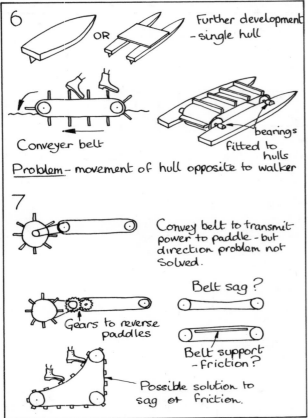

6 OR   Further development – single hull
Conveyer belt
bearings fitted to hulls
Problem – movement of hull opposite to walker

7 Convey belt to transmit power to paddle – but direction problem not solved.
Gears to reverse paddles
Belt sag?
Belt support – friction?
Possible solution to sag & friction.

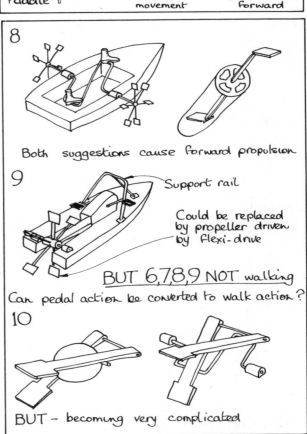

8 Both suggestions cause forward propulsion

9 Support rail
Could be replaced by propeller driven by flexi-drive

BUT 6,7,8,9 NOT walking
Can pedal action be converted to walk action?

10 BUT – becoming very complicated

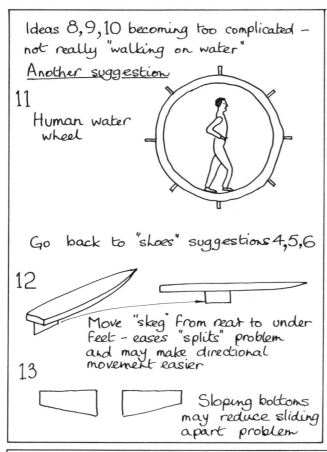

Ideas 8, 9, 10 becoming too complicated — not really "walking on water"

Another suggestion

11

Human water wheel

Go back to "shoes" suggestions 4, 5, 6

12

Move "skeg" from rear to under feet — eases "splits" problem and may make directional movement easier

13

Sloping bottoms may reduce sliding apart problem

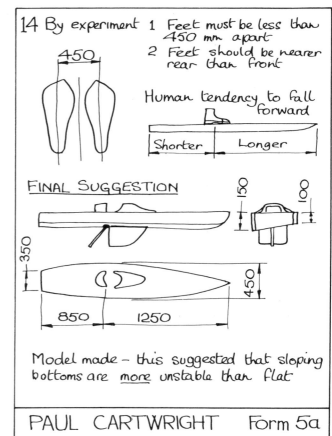

14 By experiment 1 Feet must be less than 450 mm apart
2 Feet should be nearer rear than front

450

Human tendency to fall forward

Shorter   Longer

FINAL SUGGESTION

150   100

350   450

850   1250

Model made — this suggested that sloping bottoms are more unstable than flat

PAUL CARTWRIGHT   Form 5a

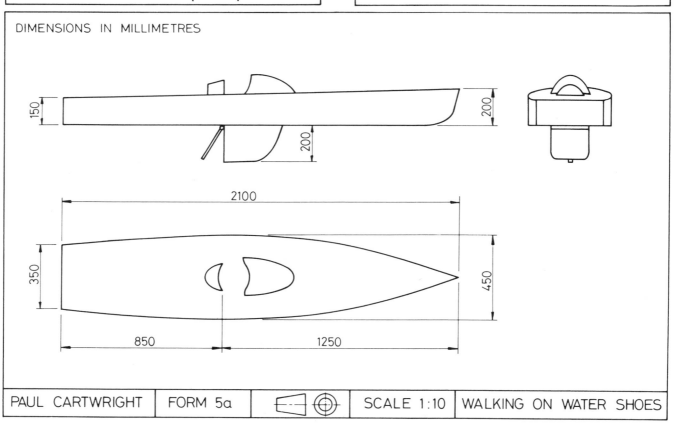

DIMENSIONS IN MILLIMETRES

150   200

200

2100

350   450

850   1250

| PAUL CARTWRIGHT | FORM 5a | | SCALE 1:10 | WALKING ON WATER SHOES |

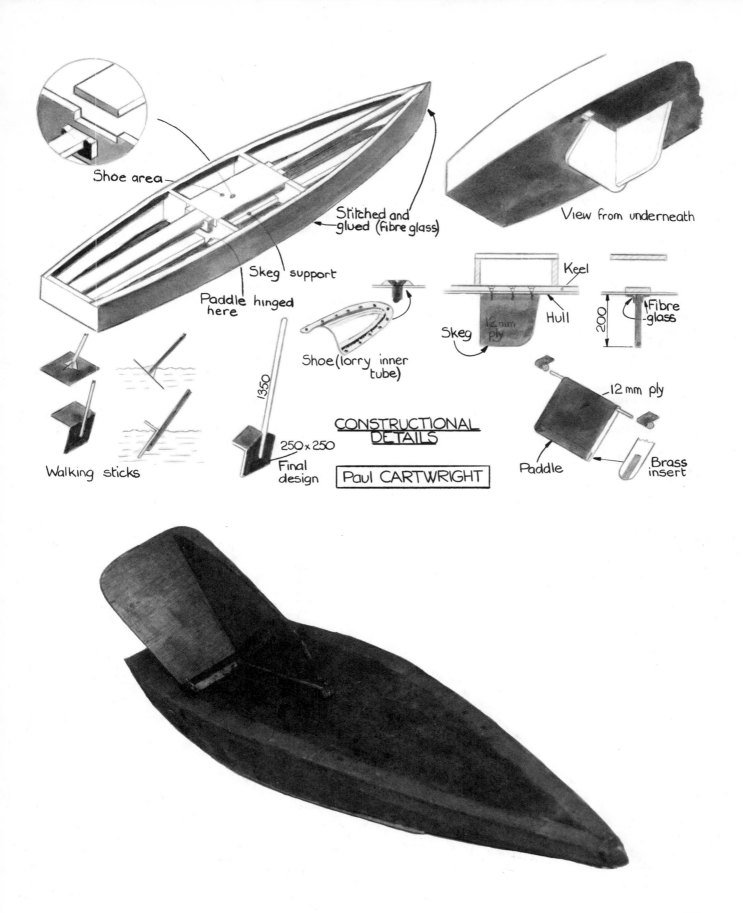

Shoe area

Stitched and glued (fibre glass)

View from underneath

Skeg support

Paddle hinged here

Shoe (lorry inner tube)

Keel

Skeg

12 mm ply

Hull

200

Fibre glass

12 mm ply

Walking sticks

1350

250 x 250

Final design

CONSTRUCTIONAL DETAILS

Paul CARTWRIGHT

Paddle

Brass insert

Project 7 Dramatics effects box

Situation

As part of the production of a stage play, a device is required which produces the following sequence of visual and aural effects:

1. Bell ringing for 10 seconds
2. Bubbles drifting across stage for 15 seconds
3. Flags waving above a box for 10 seconds
4. A repeat of 1, 2 and 3
5. A loud bang followed by a shower of confetti coming from the box.

Design brief

Design a box to stand on a table centre-stage, which produces the effects when cued by a given signal.

Analysis and investigation

Function—To provide the stage effects required by the producer of the play.

Shape and form—The device is to be made in the shape of a wooden box with three flags protruding. A series of connections to the rear or wings of the stage can be added if necessary.

Materials and equipments—The device is a stage 'prop' with a life limited by the run of the play. In the chosen design the box containing the effects equipment was made from 8 mm plywood and other pieces of equip-ment from wood offcuts, strips of 12 mm by 3 mm mild steel, pieces of tinned sheet metal, pieces of mild steel from a workshop scrap box, lengths of twine and a length of plastic piping. The electric motors and other electrical equipment were 'borrowed' from technology projects for the period during which the play was running.

Shaping, forming and jointing—Because of the temporary nature of the device, all parts were glued and pinned or screwed together.

Surface finish—Apart from painting the outside of the box containing the equipment, no other finish was thought to be necessary.

Safety—Motors running off low voltage batteries. Blank cartridge in 'bang' and confetti device.

Economics—The costs will be minimal because of offcut and scrap material used. Costs will in any case be recovered from overall profits realised by sale of tickets for the play.

Notes on drawings

The drawings are numbered 1 to 9.

Drawing 1—Two bell circuits are shown. That on the left would cause the bell to ring continuously when the switch is closed. Switching through a relay controlled through a capacitor could produce a 10 seconds ring. Switch must close momentarily just long enough to charge the capacitor.

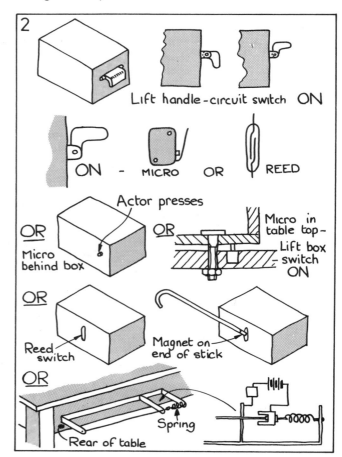

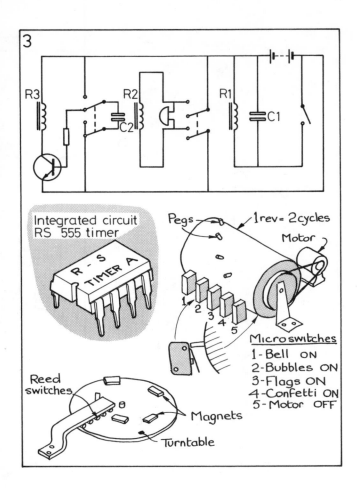

3

Integrated circuit RS 555 timer

R - S TIMER A

Pegs — 1 rev = 2 cycles

Motor

Microswitches
1-Bell ON
2-Bubbles ON
3-Flags ON
4-Confetti ON
5-Motor OFF

Reed switches

Magnets

Turntable

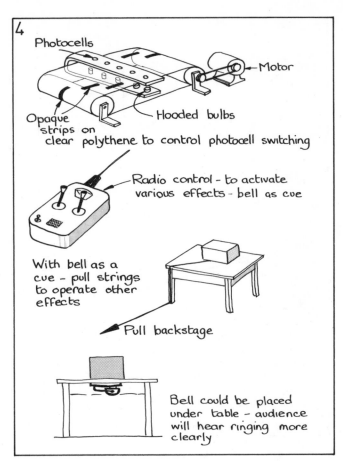

4

Photocells

Motor

Opaque strips on clear polythene to control photocell switching

Hooded bulbs

Radio control - to activate various effects - bell as cue

With bell as a cue - pull strings to operate other effects

Pull backstage

Pull

Bell could be placed under table - audience will hear ringing more clearly

Drawing 3—The electrical circuit shows a system which could be made up for controlling the effects. The circuit operates as follows:

1. Press switch and release.
2. Capacitor C1 charges; when switch released C1 discharges and relay R1 closes—bell rings and R2 closes; C2 charges.
3. C1 stops discharging; R1 opens; bell stops; R2 opens.
4. C2 discharges; circuit with transistor operates R3. Extending this suggestion could give the required sequence. It is also possible to use a 555 integrated circuit in a timing circuit to operate the effects.

Drawing 4—After considering switching the required effects by various electrical or mechanical timing arrangements, a decision was made that these were becoming too complicated for a stage 'prop'. The solution chosen for obtaining the required sequence was to use pull cords and a plastic pipe connected to an air supply operated from behind the stage. The bell was connected direct to a battery and simple switch circuit. The final control system became—

1. Cue—press bell switch for 10 seconds.
2. Pull cord to raise bubble ring, blow air through pipe, release cord to lower bubble ring and raise again as necessary for 15 seconds.

3. Pull cord to switch on motor operating flags. Release after 10 seconds.
4. Repeat 1, 2 and 3.
5. Pull cord to operate motor to pull trigger of starter gun. After bang and confetti shower, release cord.

5

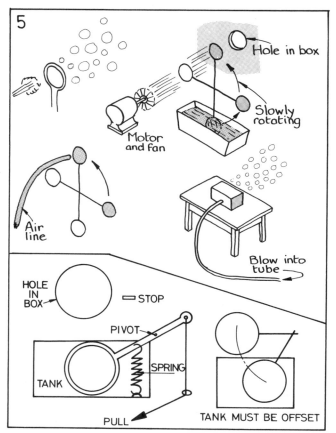

Hole in box

Slowly rotating

Motor and fan

Air line

Blow into tube

HOLE IN BOX — STOP

PIVOT

SPRING

TANK

PULL

TANK MUST BE OFFSET

6

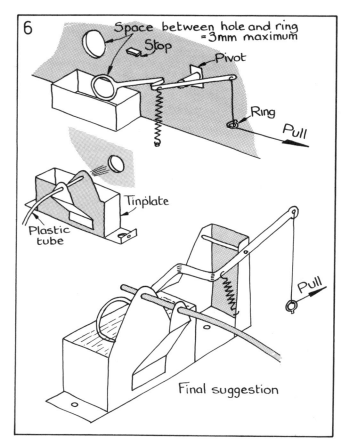

Space between hole and ring = 3mm maximum

Stop

Pivot

Ring

Pull

Tinplate

Plastic tube

Pull

Final suggestion

7

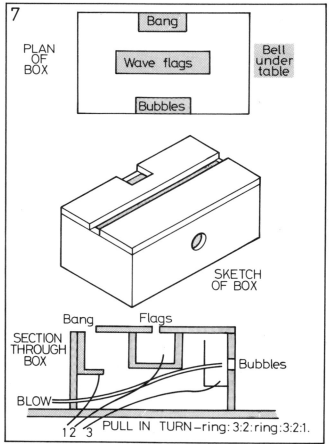

PLAN OF BOX

Bang

Wave flags

Bell under table

Bubbles

SKETCH OF BOX

Bang

Flags

SECTION THROUGH BOX

Bubbles

BLOW

1 2 3

PULL IN TURN—ring: 3:2: ring: 3:2:1.

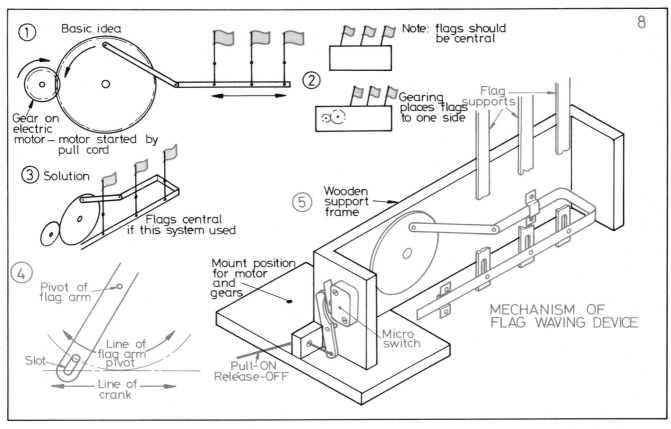

① Basic idea

Gear on electric motor — motor started by pull cord

Note: flags should be central

② Gearing places flags to one side

③ Solution

Flags central if this system used

④ Pivot of flag arm

Line of flag arm pivot

Slot

Line of crank

⑤ Wooden support frame

Flag supports

Mount position for motor and gears

Micro switch

Pull-ON Release-OFF

MECHANISM OF FLAG WAVING DEVICE

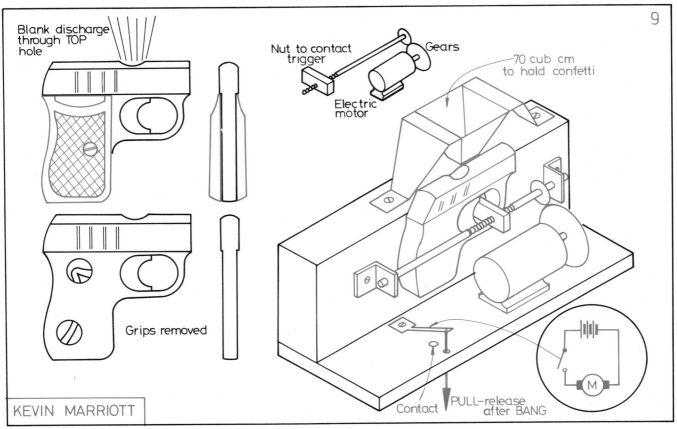

Blank discharge through TOP hole

Grips removed

Nut to contact trigger

Gears

Electric motor

70 cub cm to hold confetti

Contact

PULL-release after BANG

KEVIN MARRIOTT

Project 8 Three-minute timer

Situation
I am a keen photographer and develop and print my own films and photographs.

Design brief
Design a timing device, accurate to 1 second, which can be adjusted to give timings between 0 and 3 minutes.

Analysis and investigation
Function—An adjustable timing device which is to produce a signal when a chosen time has elapsed. The signal should be easy to see (or hear). The device needs to be easy to adjust in darkroom conditions.

Shape and form—Not of any importance but the device should be reasonably small in order to avoid taking up too much space on the darkroom's working surfaces.

Materials—Box can be made from a hardwood. Electrical circuit components can be mounted or set into transparent acrylic sheet.

Shaping and forming—The acrylic sheet will be heat formed to its required shape. A box to hold the sheet will be made from hardwood.

Jointing—Heat sink soldering required to join the electrical components. Box can be jointed as simply as possible.

Strength—The device requires to be sufficiently robust in construction to withstand the possibility of being knocked from the darkroom table on to the floor.

Surface finish—Smooth. Chemical and water resistant.

Fittings—The drawing on page 35 lists the components needed.

Safety—Low power. Stable. No 'white' light to ruin exposures when developing negatives or prints.

Economics—Not important. The device should in any case be cheaper to construct than a commercial timer purchased from a photographic shop.

Special Factors—Easy setting; clear calibration. Possibility of the device including a circuit to switch on the enlarger light source.

Notes on drawings
Capacitor discharge drawings—When switch A is closed, relay R is activated and capacitor C is charged. When A is opened, relay R continues to latch until capacitor C discharges. This electrical delay could be used as a timing device.

A 5000 μF capacitor will produce only a short delay—perhaps 4 or 5 seconds. To obtain longer delays a number of capacitors will need to be connected in parallel.

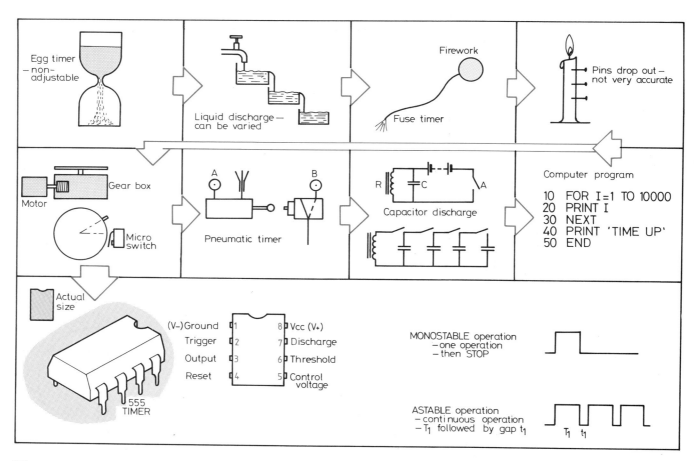

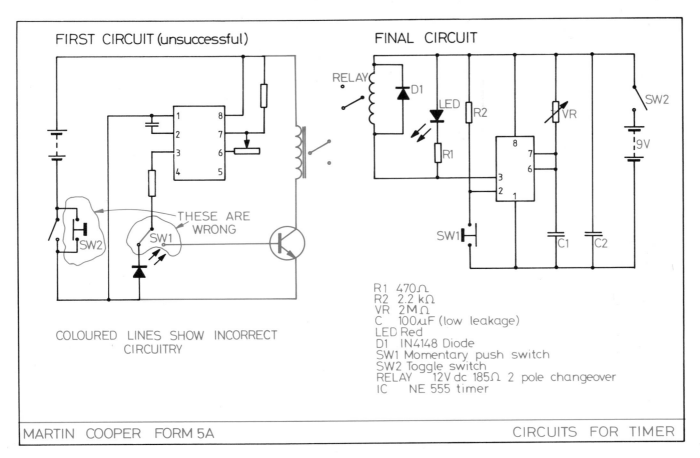

FIRST CIRCUIT (unsuccessful)

THESE ARE WRONG

COLOURED LINES SHOW INCORRECT CIRCUITRY

FINAL CIRCUIT

R1 470 Ω
R2 2.2 kΩ
VR 2 MΩ
C 100 μF (low leakage)
LED Red
D1 IN4148 Diode
SW1 Momentary push switch
SW2 Toggle switch
RELAY 12V dc 185 Ω 2 pole changeover
IC NE 555 timer

MARTIN COOPER FORM 5A   CIRCUITS FOR TIMER

Computer program

This program on Research Machines 380Z took 8 seconds to count up to 10 000, but by amending the program to give a read-out on a visual display unit (VDU) 3.04 minutes were required to count to 10 000.

The 555 Timer integrated circuit

See data sheet published by RS Components Limited. This integrated circuit timer is accurate to $\pm 2\%$ in 3 minutes. It is capable of monostable or astable operation:

   Monostable—one operation, then stop.
   Astable—continues operating T1, gap, T2, etc. The timing device requires monostable operation. The formula using the 555 timer is:

   $T = 1.1\ RAC$ seconds (RA in ohms, C in farads)

In the circuit shown time is controlled by VR and C

$$T = 1.1 \times VR \times C$$
$$= 1.1 \times 2\ M\Omega \times 100\ \mu F$$
$$= 1.1 \times 2\,000\,000 \times \frac{100}{1\,000\,000}$$
$$= 1.1 \times 2 \times 100$$
$$= 220\ \text{seconds maximum. Note VR is variable.}$$

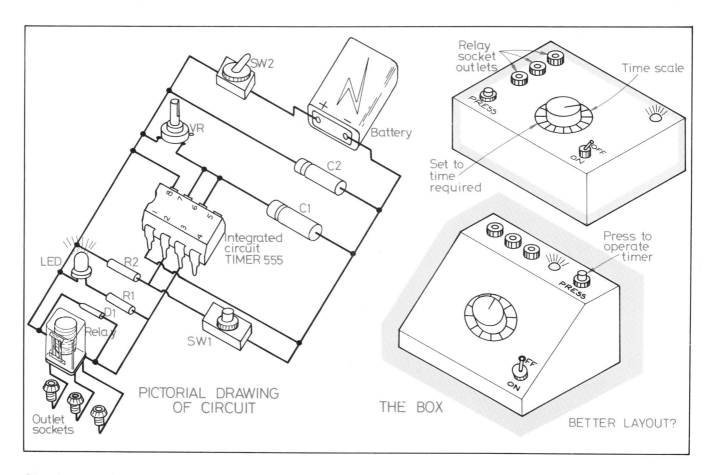

SW2

VR

C2

C1

Battery

Integrated
circuit
TIMER 555

LED

R2

R1

D1

Relay

SW1

Outlet
sockets

PICTORIAL DRAWING
OF CIRCUIT

Relay
socket
outlets

Time scale

Set to
time
required

PRESS

ON OFF

THE BOX

Press to
operate
timer

PRESS

ON OFF

BETTER LAYOUT?

Circuit operation

Turn the circuit on with SW2.

Start the timing period by pressing SW1.

Pressing SW1 causes pin 2 of the 555 timer to go below $\frac{1}{3}$ of the supply voltage. A comparator inside the integrated circuit detects this change and operates a flip-flop to make the output (pin 3) go to 9 volts. This turns off the LED and the relay. The capacitor C now starts charging via VR. The voltage across the capacitor increases as the charge in the capacitor increases. When this voltage reaches $\frac{2}{3}$ of the supply voltage, a further comparator inside the IC causes the flip-flop to re-set. This makes the output voltage at pin 3 fall to nearly 0 volts—turning on the LED, operating the relay and discharging the capacitor ready for the next delay.

Situation and Design brief

A project to design and make a model hovercraft which will attain as high a speed as possible on a control line

Analysis and Investigation

Function
Described in brief

Materials
Expanded polystyrene foam, cardboard, model wheel, 'Plasticard', wood and acrylic sheet

Shape and form
Described in solution drawings

Jointing
Mainly with adhesives - 'Araldite', balsa cement, polystyrene cement. Control box-screwed joints

Strength
The hull is of sufficient strength to enable the model to fulfil its design function

Surface finish
All surfaces to be as smooth as possible to reduce air drag

Fittings
12 Volt battery, switch, 6V motor, 12V motor, relay, resistor, connecting cable, control line cable, sockets for control box

Safety
Low voltage

Hovercraft model

Nigel Heath

POSSIBLE SOLUTIONS

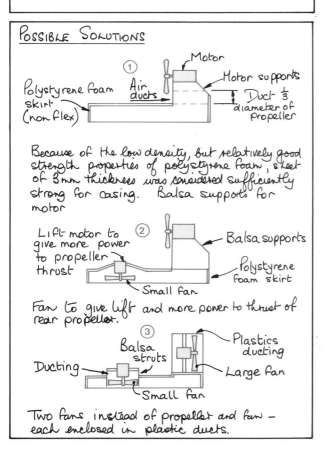

① Motor, Air ducts, Polystyrene foam skirt (non flex), Motor supports, Duct ⅓ diameter of propeller

Because of the low density, but relatively good strength properties of polystyrene foam, sheet of 3mm thickness was considered sufficiently strong for casing. Balsa supports for motor

② Lift motor to give more power to propeller thrust, Balsa supports, Polystyrene foam skirt, Small fan

Fan to give lift and more power to thrust of rear propeller.

③ Balsa struts, Plastics ducting, Ducting, Large fan, Small fan

Two fans instead of propeller and fan — each enclosed in plastic ducts.

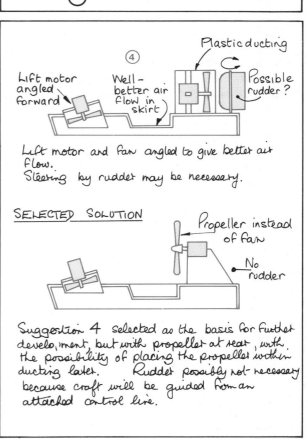

④ Plastic ducting, Lift motor angled forward, Well - better air flow in skirt, Possible rudder?

Lift motor and fan angled to give better air flow.
Steering by rudder may be necessary.

SELECTED SOLUTION

Propeller instead of fan, No rudder

Suggestion 4 selected as the basis for further development, but with propeller at rear, with the possibility of placing the propeller within ducting later. Rudder possibly not necessary because craft will be guided from an attached control line.

HULL

Hull to be constructed from 5mm polystyrene foam tiles - sufficiently light + strong. All joints glued with pva glue - obviates nails etc.

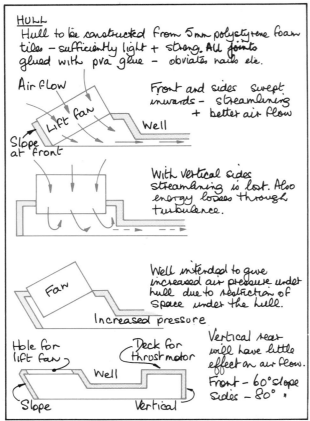

Air flow

Lift fan

Slope at front

Well

Front and sides swept inwards - streamlining + better air flow

With vertical sides streamlining is lost. Also energy losses through turbulence.

Fan

Increased pressure

Well intended to give increased air pressure under hull due to restriction of space under the hull.

Hole for lift fan

Well

Deck for thrust motor

Slope

Vertical

Vertical rear will have little effect on air flow. Front - 60° slope Sides - 80° "

FAN MOUNTING

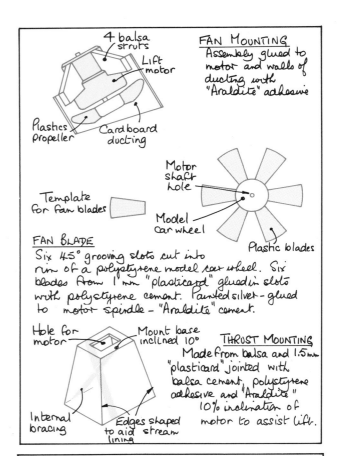

4 balsa struts

Lift motor

Assembly glued to motor and walls of ducting with "Araldite" adhesive

Plastic propeller

Cardboard ducting

Template for fan blades

Motor shaft hole

Model car wheel

Plastic blades

FAN BLADE

Six 45° grooving slots cut into rim of a polystyrene model car wheel. Six blades from 1mm "plasticard" glued in slots with polystyrene cement. Painted silver - glued to motor spindle - "Araldite" cement.

Hole for motor

Mount base inclined 10°

Internal bracing

Edges shaped to aid stream lining

THRUST MOUNTING

Made from balsa and 1.5mm "plasticard" jointed with balsa cement, polystyrene adhesive and "Araldite" 10% inclination of motor to assist lift.

ELECTRIC CIRCUIT.

Two D.C. motors are available - a 12V and a 6V. The 12V motor, capable of 2000 rev/min is suitable for driving the thrust propeller. It is best controlled by a double-pole switch, which will provide a reversing system in the circuit for the propeller.

The 6V motor is therefore to be used to drive the lift fan. This motor requires 0.6A to operate. Its resistance must therefore be by Ohm's Law

$6 \div 0.6 = 10\,\Omega$

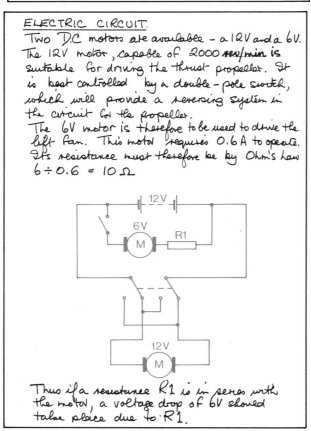

12V

6V

M

R1

12V

M

Thus if a resistance R1 is in series with the motor, a voltage drop of 6V should take place due to R1.

CONTROL BOX

A control box is to be made to house the electric circuitry

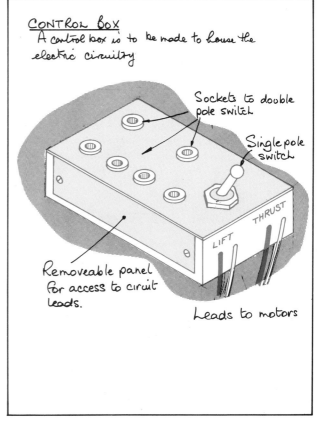

Sockets to double pole switch

Single pole switch

Removeable panel for access to circuit leads.

LIFT

THRUST

Leads to motors

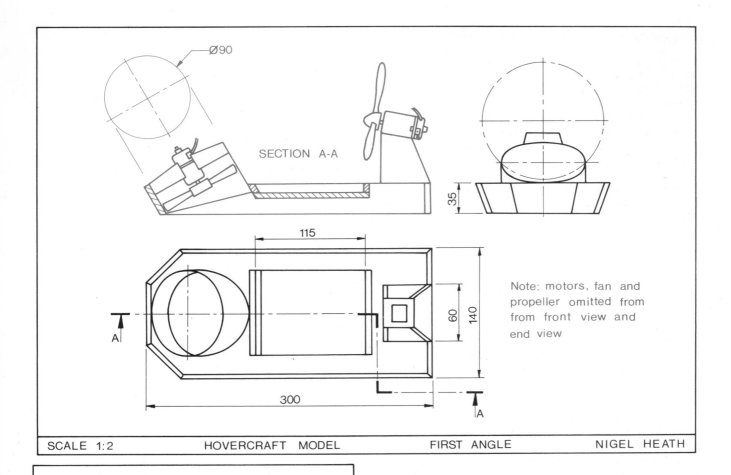

Ø90

SECTION A-A

Note: motors, fan and propeller omitted from from front view and end view

35

115

60

140

300

A

A

MODIFICATIONS

Some modifications were carried out on the original hovercraft as a result of experiments. The most important of these modifications were:-

1. A larger motor and larger fan and duct was fitted to give increased and better lift.

2. The well depth was decreased to give more space underneath the hull

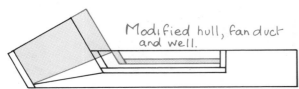

Modified hull, fan duct and well.

Original hull and fan duct.

3 By removing some of the fan duct a greater under-hull air pressure was obtained.

APPRAISAL

After the modifications detailed above, the craft performed well. In particular the use of adhesives for joining the various parts proved to be particularly successful.

Photograph by courtesy of British Petroleum

Project 10 Aero-generator

Details of this aerogenerator project were made available by permission of the staff of the Walton High School, Stafford, where the project was designed and made. The authors would like to place on record here their appreciation of the generous help given, in both passing on information and allowing photographs to be taken, by the head teacher of the school, J. E. Wilkinson, Esq., M.A., and to N. A. Brittain, Esq., the teacher responsible for the project. The following pupils took part in the project—David Mellhuish, Robert Howells, Stephen McCosh, Mark Edwards, Martin Ovelton, Andrew Booth, Jeremy Robinson, Trevor Powell and David Proughton.

Situation
A greenhouse (growing unit) requires a heating system.

Design brief
Design a heating system for a greenhouse of average size (4.5 m by 2.5 m) capable of maintaining a frostfree environment at minimum energy costs.

Analysis and investigation
Function—To create heat and preserve the heat when created with a minimal effect on the greenhouse facilities.
Shape and form—Not of particular importance, but the design should occupy as little space as possible, preferably parts of the greenhouse not used for growing plants.
Materials—All materials employed must be durable and weather proof. Care should be taken to avoid materials which may affect plant growth.
Jointing—Watertight and weatherproof electrical insulation. Jointing should be able to withstand accidents—such as dropped plant pots, movement of soil, knocking when using watering cans etc.
Strength—System must be sufficiently robust to withstand damage by inexpert handling.
Fittings—Best to place an 'over-design' emphasis when choosing fittings. Reliability of the system is very important.
Safety—Slight fire risk. Possible damage of propeller mounting and bearings at high wind velocity—hence high propeller speeds. Tower construction for the propeller and generator unit must be capable of withstanding high winds. Protect the tower against children climbing.
Economics—Much of the equipment used was second-hand of mainly scrap value. Running costs are small.

Power sources which were considered
1. Some form of waste burning heater/boiler
Wood—difficult to obtain in quantity.
Dustbin waste—difficult to make combustible.
Sump oil—possible—oildrum/water drip burners?

2. Heat generated by bio-chemical actions
Compost decomposition.
Manure—methane gas.
3. Sun power
Insulation—greenhouse hot during daytime.
Sun heated storage units—bricks or water.
Solar cell for electricity/heat conversion.
Solar panel for water heating.
4. Water
No natural or free standing source available.
5. Wind power
Possible electric or rotary pumping action—site is open with little air turbulence.

Pump control
The pump should preferably only circulate water through the system when the solar panel is heating the water—not at night or when sunlight is inadequate during the day.

The determining factor is the difference in temperature between water entering the solar panel and water leaving the panel.

Temperature IN (A)	Temperature OUT (B)	Pump
Lower than B	Higher than A	ON
Same as B	Same as A	OFF
Higher than B	Lower than A	OFF

A solar panel for water heating

40

POWER SOURCES TO DRIVE PUMP

1

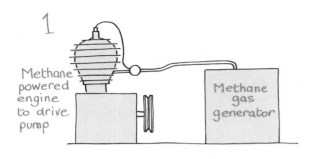

Methane powered engine to drive pump

Methane gas generator

2

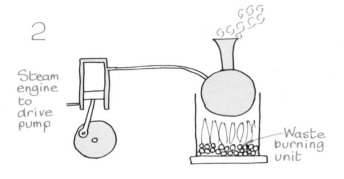

Steam engine to drive pump

Waste burning unit

3

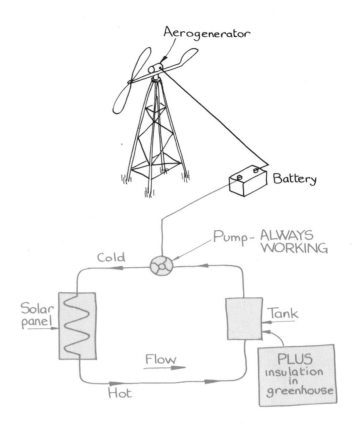

Wind driven pump

Aerogenerator

Battery

Pump- ALWAYS WORKING

Cold

Solar panel

Flow

Hot

Tank

PLUS insulation in greenhouse

4

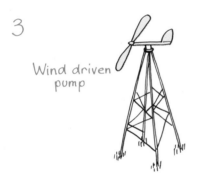

Electrically driven pump

Battery

Best solution — BUT electricity is an expensive power source

A photograph of the control circuitry shown on page 43

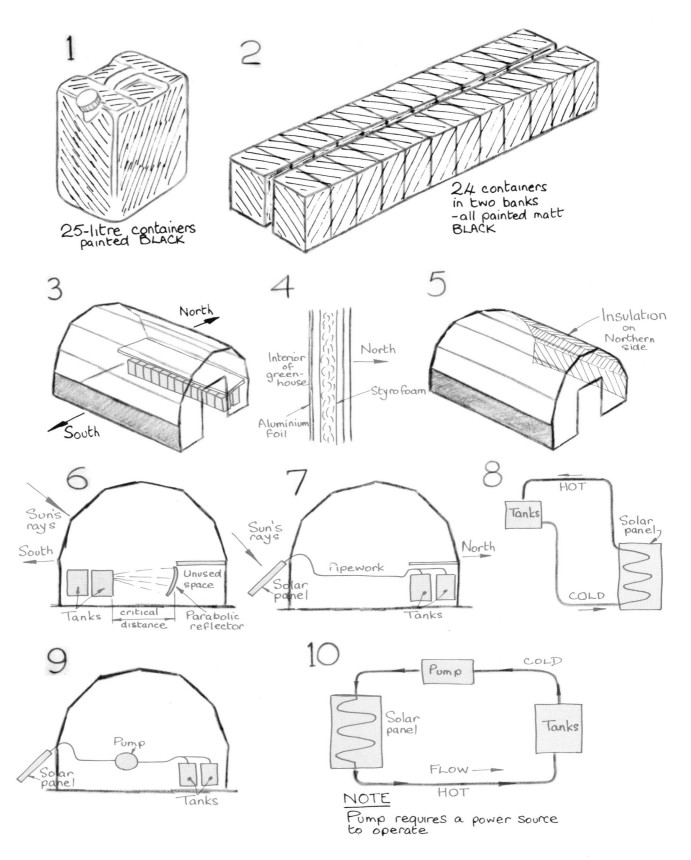

1 25-litre containers painted BLACK

2 24 containers in two banks -all painted matt BLACK

3 North / South

4 Interior of greenhouse / North / Styrofoam / Aluminium foil

5 Insulation on Northern side

6 Sun's rays / South / Tanks / critical distance / Unused space / Parabolic reflector

7 Sun's rays / Pipework / North / Solar panel / Tanks

8 HOT / Tanks / Solar panel / COLD

9 Pump / Solar panel / Tanks

10 Pump / COLD / Solar panel / Tanks / FLOW → / HOT

NOTE
Pump requires a power source to operate.

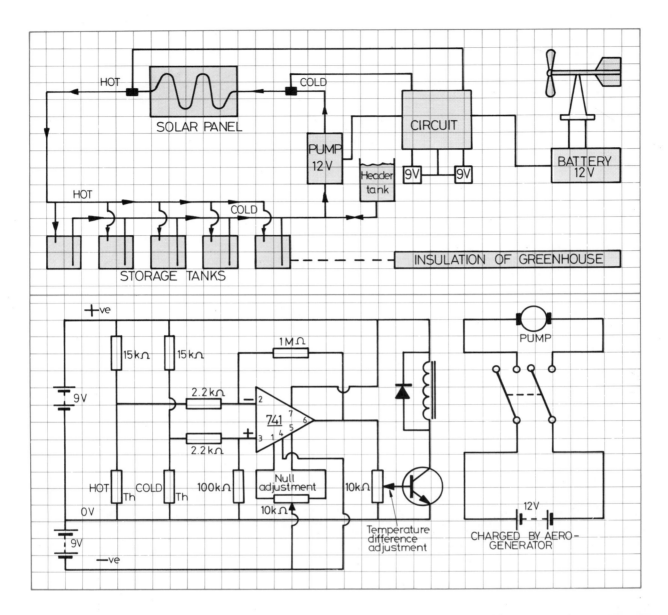

Solar panel control circuit

1. The wind driven generator constantly charges the battery when the propeller is rotating.
2. The battery (12 volt) supplies power through the circuit via a relay operated switch to the pump.
3. The circuit takes its power from two 9 v batteries.

The circuit

The circuit has two inputs: one from the 'Hot' thermistor and the other from the 'Cold' thermistor. The only output of the circuit is to the pump and this is switched on when the temperature difference between the two thermistors exceeds a preset level. The thermistors are mounted each side of the solar panel, the 'Cold' some distance from the input side and the 'Hot' very close to the output side. The difference in the thermistor signals represents the temperature rise of the water due to solar heating. If both thermistors were close to the solar panel then without the pump working, no temperature difference would register and therefore the circuit would never switch the pump on.

The pump is thus only on when sunlight on the solar panel heats the water by a certain amount. The amount can be varied by altering the 10 kΩ 'temperature difference adjustment' potentiometer.

The 741 integrated circuit chip

The heart of this circuit is the 741 operational amplifier integrated circuit chip.

The 741 operational amplifier is a high gain amplifier which will amplify any difference in the two input voltages from the thermistors. The gain on a 741 may be 10 000 or more—but this can be reduced. The two inputs are at pins 2 and 3. Any difference in these two voltages is amplified and the resulting output fed into pin 6. See also page 93.

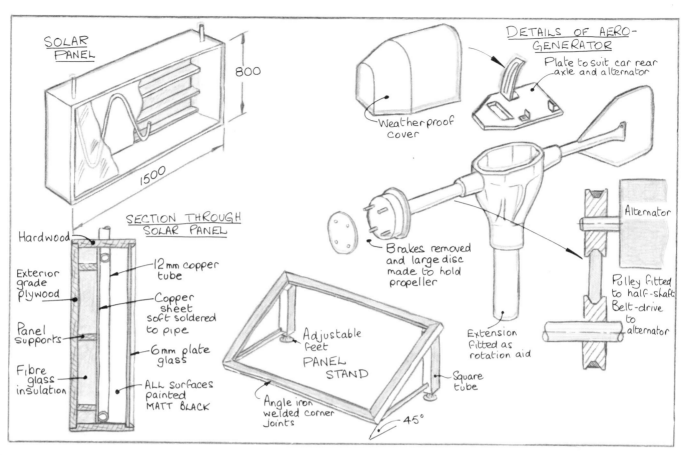

SOLAR PANEL

800

1500

DETAILS OF AERO-GENERATOR

Plate to suit car rear axle and alternator

Weatherproof cover

SECTION THROUGH SOLAR PANEL

Hardwood

Exterior grade plywood

Panel supports

Fibre glass insulation

12 mm copper tube

Copper sheet soft soldered to pipe

6 mm plate glass

ALL surfaces painted MATT BLACK

Brakes removed and large disc made to hold propeller

Extension fitted as rotation aid

Alternator

Pulley fitted to half-shaft. Belt-drive to alternator

Adjustable feet PANEL STAND

Angle iron welded corner joints

Square tube

45°

Use of colour in graphics

A variety of methods of presenting the graphics needed when designing technology projects are shown on pages 11 to 44. Some of these include colour. On this page and pages 46, 47 and 48 further examples of graphical presentation involving colour are considered. These examples have been copied from design sheets and folders which have been drawn as part of the design and description of technology projects in schools. Further examples of methods of presentation are given in Section 4, commencing on page 146. Colour can be applied with crayons, by paint brush and water colours, with coloured pens and, in some cases, by pasting coloured papers in position.

Drawing 1—Line drawings on a square grid. Square papers—available on 2 mm, 5 mm and 10 mm grids—enable line drawings to be produced rapidly. Freehand or instrument drawings are suitable when working on square grids. The two drawings show equipment for converting the rotation of a wheel into electrical energy and a line shaft operating two jack lifts.

Drawing 2—A line drawing on an isometric grid. Isometric grid papers—available on 5 mm, 10 mm and 20 mm grids—enable pictorial drawings to be rapidly made, either freehand or with instruments.

Drawing 3—A freehand drawing in which a coloured line has been drawn inside a black outline. Such 'in-line' drawing can attract attention to an item of graphics which is of importance in a design.

Drawing 4—Heavy background outlining. The outline of the drawing has been followed with a broad band of colour.

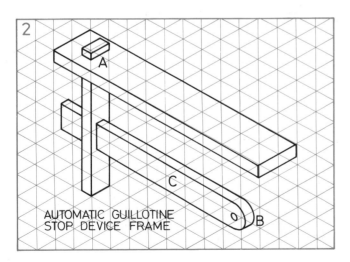

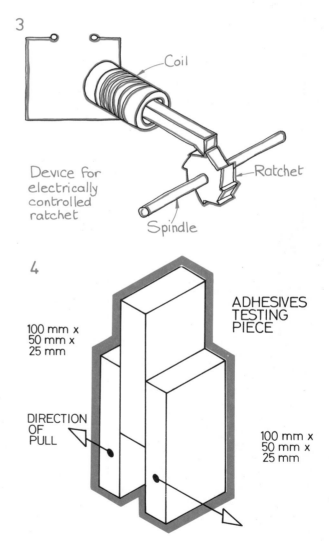

5

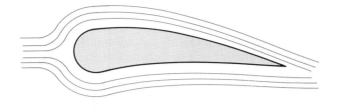

6

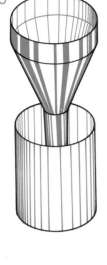

7

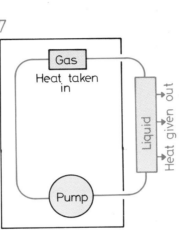

Drawing 5—A clear wash—light crayon shading, or a water colour wash applied by brush—draws attention to an important shape. This drawing shows a section through an aerofoil wing.

Drawing 6—Shading of cylindrical or conical parts with lines or bands of colour—pen or crayon.

Drawing 7—Different depths of shade of the same colour and outlining of parts of a circuit with coloured lines. The drawing shows a refrigeration circuit in diagrammatic form.

Drawing 8—A method of showing a progression of ideas when designing. Short arrows, together with arrows on broad bands show the stages by which one idea may lead to another and eventually to a reasonable solution of a technological problem. In this example the drawings show suggestions for controlling the steering of a remotely controlled vehicle.

8

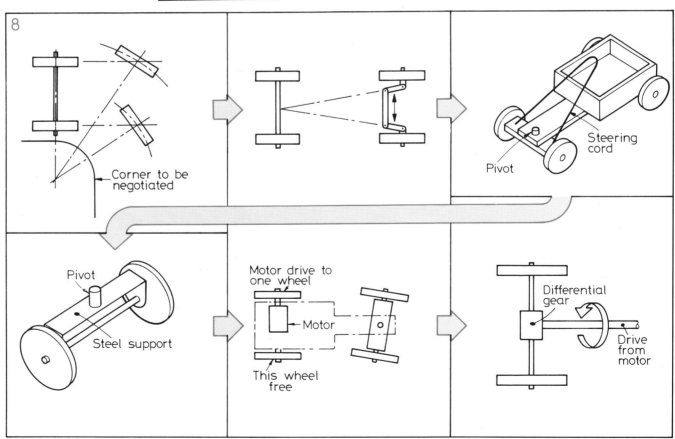

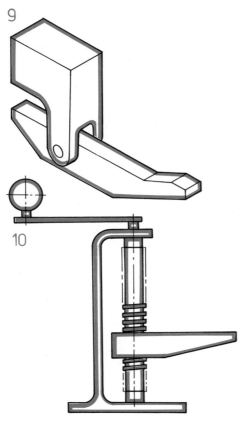

9

10

Drawings 9 and 10—Two drawings showing in-line emphasis by coloured lines following the outline of the original black line drawing. Both these examples have been drawn with the aid of instruments. Drawing 9 is part of a lifting device to be operated hydraulically. Drawing 10 is another lifting device to be operated manually.

Drawing 11—Emphasis given by coloured lines to the essential parts of an outline drawing of part of a prototype precipitation apparatus. The project involved an attempt to solve a smoke pollution problem by cleansing air passing through the precipitation apparatus.

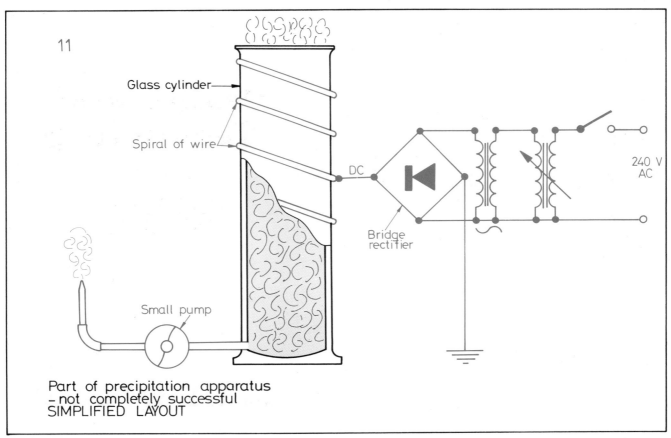

11

Glass cylinder

Spiral of wire

DC

Bridge rectifier

240 V AC

Small pump

Part of precipitation apparatus
—not completely successful
SIMPLIFIED LAYOUT

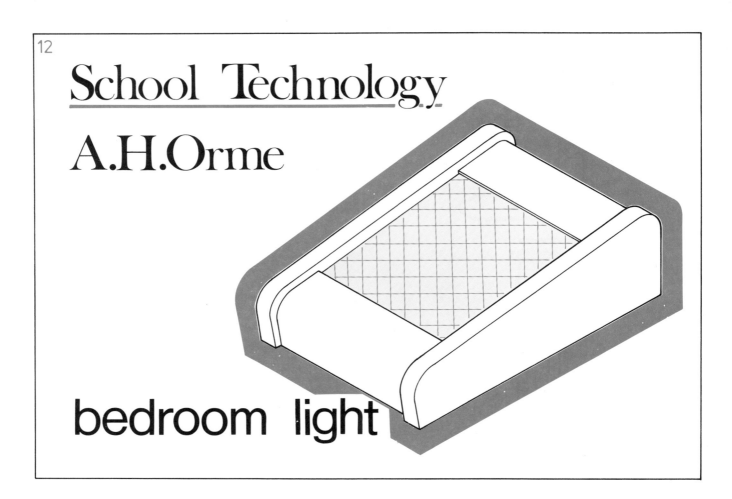

12

School Technology

A.H.Orme

bedroom light

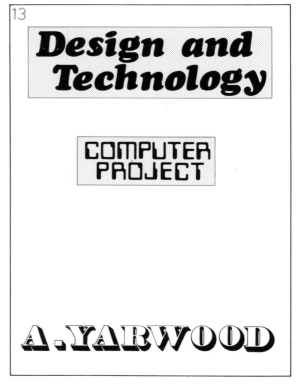

13

Design and Technology

COMPUTER PROJECT

A.YARWOOD

Drawings 12 and 13—Two drawings showing designs for folder covers. The first involves a pictorial drawing of the completed project—the bedroom light described on pages 14 to 17. The second consists of printing applied with Letraset dry transfer print, with added lines in black and colour to emphasise the lettering.

SECTION 2

Areas of Study

Introduction

Suggestions for elementary work from five areas of study are given in this section. Some knowledge in each of these areas may be necessary when designing a technology project. These areas of study are—Structures; Mechanisms; Electric circuits; Electronics; Computers. It must be emphasised that the suggestions given in this section only form an elementary basis from which the student can extend his studies by further work and reference to other sources—books, experimentation, films, slides, television, radio etc. It should also be emphasised that these five areas of study can not fully complete the background of knowledge necessary to the potential technology project designer. Some projects may need knowledge derived from other areas, as varied as—Energy and its sources; Control systems; Pneumatics and Hydraulics; Optics; Acoustics, Aerodynamics—to mention some. The method of study suggested in this section is a largely experimental one. Suggestions for practical work are offered from which the reader can make deductions. Some exercises are given associated with the practical work. Many of the suggestions and exercises could be regarded as mini technology projects.

Units of measurement

SI units of measurement are used throughout this book. Standard SI abbreviations are also employed. All SI units are based on the metric system and the units employed may be used with prefixes showing multiples or fractions of the unit—as follows:

Prefix	Value	Symbol
mega	one million	M
kilo	one thousand	k
hecto	one hundred	h
deca	ten	da
deci	one tenth	d
centi	one hundredth	c
milli	one thousandth	m
micro	one millionth	μ
nano	one thousand millionth	n
pico	one million millionth	p

Forces

Isaac Newton's Third Law states Reaction is always opposite and equal to Action, or the actions of two bodies mutually on one another are always equal and tend in opposite directions. If a weight of mass 1 kilogram (kg) rests on a floor, it is exerting 1 kilogram force (kgf) vertically downwards on the floor. In order that the mass does not fall vertically downwards, the floor must be exerting an equal and opposite 1 kgf vertically upwards on the mass of 1 kg.

1 kilogram force (1 kgf) is approximately equal to 9.81 newtons (9.81 N). This value varies slightly depending on the position in the world where the value is taken. For the purposes of the work suggested in this section, a force of 1 kgf is taken as being equal to 10 newtons.

The following are the units employed in this book.

Quantity	Unit	Symbol
length	metre	m
velocity	metre per second	m/s
acceleration	metre per second per second	m/s^2
mass	gram	g
	kilogram	kg
density	grams per cubic centimetre	g/cm^3
force	newton	N
moment	newton metre	N m
	newton millimetre	N mm
electrical current	ampere	A
potential difference	volt	V
resistance	ohm	Ω
capacitance	farad	F
work	joule	J
power	watt	W
angle	degree	°
	minute	'
	second	"
	radian	rad

Structures

Structures need to be designed so that they are able to resist and withstand forces which may cause them to become distorted. Forces applied to structures may cause:
1. Bending.
2. Buckling.
3. Flexing.
4. Stretching.
5. Shearing.
6. Twisting.

Some methods by which structures can be designed so as to reduce the possibility of distortion are shown in a series of drawings on this page and on page 51 opposite.

1. Forces tending to cause bending
Such forces can be resisted by means of:
(a) Adding a central support, or supports, across the material.
(b) Placing materials on edge so that the widest part of the material takes the impact of the force.
(c) Adding central supports along the length of the material.
(d) Bending the edges of sheet material—in effect, supports along the edges of the material.
(e) Using angled materials—similar to an edge support.
(f) Using angled materials within which holes have been drilled to reduce weight or provide fixing or jointing holes for rivets and/or bolts.
(g) Using T shaped girders.
(h) Using H shaped girders.

2. Forces which tend to cause buckling
Such forces are compression forces. Compression forces may be resisted by means of:
(a) Sheer mass.
(b) Mass reduced to pipe, tube or U shape section. While having a similar result to mass, such sections give lightness to a structure without appreciably weakening resistance to compression.
(c) Spreading the compression forces by means of a pyramid.
(d) Using pylons, which are in effect pyramids made from small section material.

3. Forces which tend to cause flexing
Such forces may be resisted by means of:
a) Brackets.
b) Compression rods.
c) Tension ties, either vertical or at an angle.
d) Tension stays.
e) Compression supports.

4. Forces tending to cause stretching
Tension forces can generally be resisted by the use of materials which are themselves tension resistant.

The applied force tends to cause bending

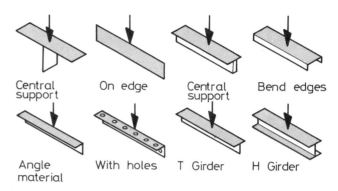

Central support   On edge   Central support   Bend edges

Angle material   With holes   T Girder   H Girder

Some remedies to counteract bending forces

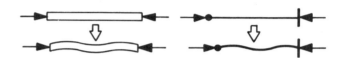

Compression forces tend to cause buckling

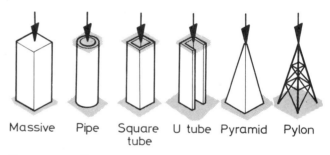

Massive   Pipe   Square tube   U tube   Pyramid   Pylon

Some remedies to counteract compression forces

The applied force tends to cause flexing

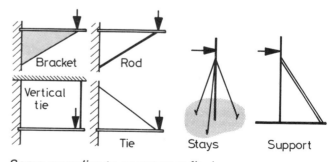

Bracket   Rod

Vertical tie   Tie   Stays   Support

Some remedies to counteract flexing

5. Forces tending to cause shearing

Care needs to be taken when designing some structures to ensure that forces will not be placed on parts of the structure in such a manner as to cause the material from which it is made to shear. The positioning of bolts and rivets, for example, needs careful thought because such jointing components can themselves shear if placed in the wrong positions.

6. Forces tending to cause twisting

Materials undergoing torsion forces or torque may become twisted. Some materials are deliberately designed to withstand such torsion forces, an example being torsion rods which are included as part of the suspension system of some road vehicles.

Examples of structures designed for maximum strength

Structures which include triangular parts tend to be strong structures. Some examples are shown. A suspension bridge is a strong structure due to the parabolic curve of the bridge cable supports. A roof truss, based on an outer triangle with braces forms a strong structure. Some exercises in determining the forces in such structures are given in the pages which follow.

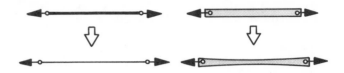

When under tension, materials tend to stretch

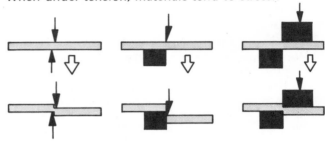

When under shear forces, materials tend to separate

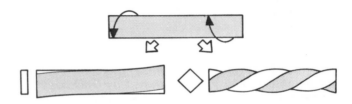

When under torsion forces, materials tend to twist

An unused test piece and a square bar which has been subjected to torsion testing in a test machine

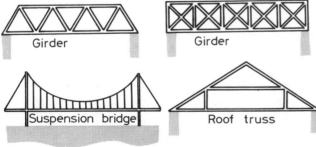

Girder   Girder

Suspension bridge   Roof truss

Some examples of structures designed for maximum strength

Forces in equilibrium

1. Make the apparatus shown in Drawing 1.
2. (a) Hang a 200 gram weight at A. What happens?
   (b) Hang a 400 gram weight at A. What happens?
   (c) Remove screw B. Hang a 200 g weight at A. What happens? Repeat with a 400 g weight.
3. (a) Remove the weight. Replace screw B. Add brace to the bracket—screwed at C and bolted at A.
   (b) Hang a 200 g weight at A. What happens? Hang a much heavier weight at A—say 2 kilograms. What happens?

Forces acting on the bracket

1. With screw B in position, the weight hung at A causes the bracket arm to deflect downwards.
2. With screw B removed, the bracket upright tends to move away from its wooden support, when weights are hung at A.
3. When the brace is added to the bracket and screw B replaced, the bracket remains rigid even when quite heavy weights are hung at A.

The three forces acting at A, which are keeping the bracket rigid can be shown diagrammatically as in Drawing 4.

Triangle of forces

The three forces of Drawing 4 can be represented by the triangle, drawn to scale, shown by Drawing 5.

Draw ab to a scale which represents the weight hung at A of Drawing 3 as a force in newtons. Take 1 kilogram force as equal to 10 newtons. Work to a scale of 20 mm represents 1 newton. If the weight hung at A is 200 grams, this can be regarded as a force of 2 newtons. At a scale of 20 mm = 1 N, the side ab of the triangle will be 40 mm. Angle bac is a right angle. Draw bc at the same angle as the brace in the bracket. In Drawings 4 and 5 this angle has been taken as being 40°.

By measurement along the sides ac and bc of the triangle of Drawing 5 the forces acting along the arm and brace of the bracket are ac (the arm) = 1.7 N; bc (the brace) = 2.65 N. The actual dimensions are ac = 3.4 mm and bc = 5.3 mm. From the scale of 20 mm = 1 N these convert to the forces given. Note these measurements are approximate because of inaccuracies inherent in drawing the triangle.

Exercises

With the brace at an angle of 35° to the upright of the bracket, draw triangles of force for the bracket when the following weights are hung at A:
1. 500 grams; 2. 300 grams; 3. 250 grams. Take 1 kilogram force as equal to 10 newtons and work to a scale of 25 mm as representing 1 newton.

Measure the lengths of the sides of the three triangles you have drawn and state the forces acting along the arm and brace in each case. *Note*—When drawing a triangle of force:

1. The three forces act in the same direction—clockwise or anti-clockwise around the triangle.
2. The triangle is complete—all angles meet.
3. The scale used to draw the triangle is a force scale, not a dimension scale.

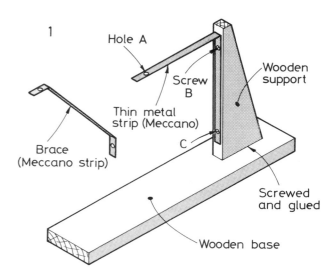

1
Hole A
Screw B
Thin metal strip (Meccano)
Brace (Meccano strip)
Wooden support
C
Screwed and glued
Wooden base

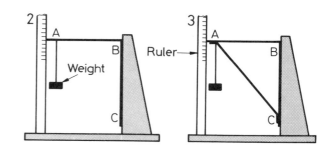

2
A
B
Weight
C
Ruler→
3
A
B
C

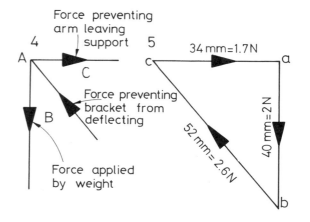

4
Force preventing arm leaving support
A
C
B
Force applied by weight

5
Force preventing bracket from deflecting
34 mm = 1.7 N
c
a
52 mm = 2.6 N
40 mm = 2 N
b

Pulleys and triangles of forces

1. Set up the apparatus—Drawing 1. The backboard can be of any size depending on the materials you have at hand. The strings must run freely in their pulleys. Place weights of 30 grams, 50 grams and 40 grams on the weight carriers as shown. Taking 1 kilogram force as equal to 10 newtons, these three weights can be regarded as forces of 3 N, 5 N and 4 N acting vertically downwards. The forces of 3 N and 4 N also act at angles along strings ba and bc as shown by arrows in the drawing. Thus at b three forces are acting—5 N vertically downwards; 3 N and 4 N acting at angles to the vertical.

2. Draw the space diagram Drawing 2. Note the lettering in the spaces of the diagram. This form of lettering is called Bow's notation. The vertical force of 5 N is known as force *AB*; that of 4 N as force *BC*; that of 3 N as force *CA*. Forces *BC* and *CA* should be drawn at the angles at which strings bc and ca rest.

3. From the space diagram draw the triangle of forces Drawing 3. Work to a scale of 10 mm represents 1 N. In this triangle ab, bc and ca are parallel to forces *AB*, *BC* and *CA*.

4. Measure the lengths of bc and ca. Measure the angles abc and bca. The lengths should correspond (in scale) to the forces of 4 N and 3 N placed on the weight carriers of the apparatus.

Note—When the weights on the apparatus come to rest, the system is said to be in equilibrium and the point b is said to be held in equilibrium by the three forces acting at b—in this case 5 N vertically down; 4 N and 3 N at angles to the vertical.

5. Repeat the experiment with weights as follows:
(a) 40 grams; 60 grams; 50 grams
(b) 50 grams; 80 grams; 60 grams
(c) 0.5 kg; 1.3 kg; 1.2 kg
(d) 1.6 kg; 2 kg; 0.9 kg.

Exercises

Draw the triangles of forces for the space diagrams shown in Drawings 4a, b, c and d. The answer to 4a has been drawn for you. Work to a scale of 10 mm represents 1 newton and assume 1 kgf = 10 N.

To find a resultant

A block of material just moves when pulled over a rough surface by two forces of 25 N and 40 N applied to two strings as shown in Drawing 5. Drawing 6 shows how to find the direction in which the block moves. The friction between the block and the surface over which it is being pulled acts as a force opposing the 25 N and 40 N forces. This frictional force can be calculated by measuring and scaling the diagonal of the parallelogram of Drawing 6. The resulting force and direction is known as the *resultant* of the two forces of 25 N and 40 N. The resultant shows the *direction* and *magnitude* of the action of the 25 N and 40 N forces. The frictional force between the block and the surface is equal to but *opposite* in direction to this resultant.

Exercises

Find, by drawing parallelograms of forces, the resultants of the forces acting at points as shown in drawings 7a, b, c and d. Work to a scale of 1 mm represents 1 N.

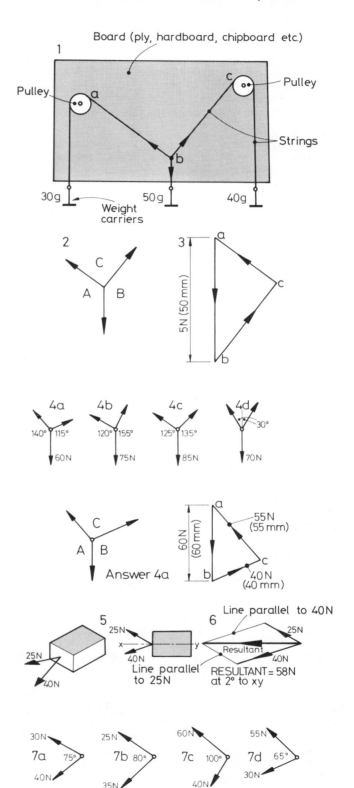

Simple frames

1. Set up the frame shown in Drawing 1. Any of a variety of materials may be used—Meccano strip, wood strips, light metal strips, strips of a rigid plastics material. Bolts, screws or nails (pins) can be used at the joints of the strips. It is important that the joints should be such that the strips can freely and easily rotate about each other at the joints. They are pin-joints. The frame should be suspended to a fixed point at A. A weight of 500 grams is suitable to act as a force of 5 newtons acting at the lower of the pin joints vertically downwards. What happens when the force is allowed to act on the frame?

2. Set up frame 2 (shown in the photograph) by adding a cross member to frame 1. What happens when the force of the weight is allowed to act on the frame?

3. Set up frame 3 held at A and B. Be careful to allow the pin joints at A and B to rotate if necessary. What happens when the force of the weight is allowed to act on the frame?

4. Add a diagonal strip to frame 3. Apply the force as before. What happens to the frame?

5. Set up frame 5. The frame should be firmly held at A and B while allowing the pin joints at those points to rotate. What happens when the weight is allowed to act as a force on the frame?

6. Add a diagonal strip to frame 5. Apply the force of the weight. What happens?

7. Set up frame 7 held at A and B. Let the force act on the frame. What happens? How can frame 7 be made rigid?

8. Set up frame 8 held at A and B. Apply the force of the weight. What happens?

9. Add a second diagonal strip to frame B. Apply the force. What happens?

10. Make up frame 10 held at A and B. Apply the force of the weight, first as shown, then at C. What happens? How can this frame be made rigid?

11. Set up frame 11 held at A and B. Apply the force of the weight. What happens? How can this frame be made fully rigid?

Note—In order to make pin-jointed frames *simply firm* the rule is—number of bars = (2 × number of joints)

54

minus 3. Check this. Additional bars or struts are *redundant* members for such a frame.

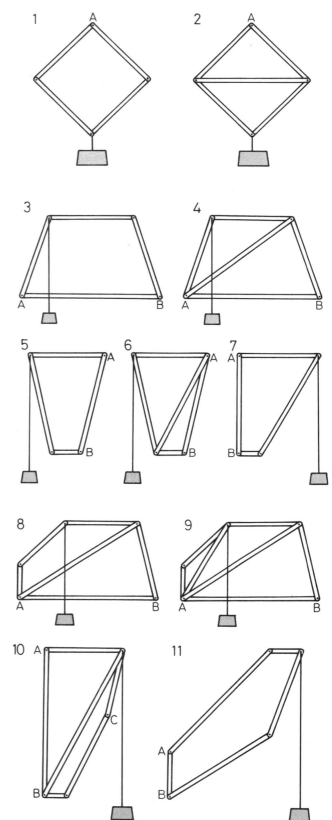

Polygon of forces

The theory behind the triangle of forces can be carried a stage further in cases where more than three forces act at a point. In such cases the forces involved may be calculated by drawing a *polygon of forces*. The same rules governing the triangle of forces apply to polygons of forces.

Drawing 1—The point O is held stationary by four forces, only two of which are known. To find the magnitude of the other two forces draw the polygon *Drawing 2*—this is a quadrilateral because four forces are involved.

1. Draw line A parallel to force *A* and of a scaled length say 10 mm represents 1 newton i.e. 50 mm long.
2. From one end of this line draw a second line parallel to force *B*, to the same scale i.e. 60 mm long.
3. Complete the polygon by drawing the other two lines parallel to forces *C* and *D*, from the ends of lines A and B.
4. The magnitude of forces *C* and *D* can now be calculated from the scaled lengths of the two lines in the polygon.

Note—If the point O is to be in equilibrium:
1. The polygon must be complete—it must close.
2. The directions of the arrows of force must all flow in the same sense around the polygon.
3. The polygon may be drawn in two ways to obtain the same result—Drawings 2 and 3 are congruent.

To find a resultant of more than two forces
Specimen Question
Drawing 4—To find a single force which will replace the three forces acting at O as shown.
Specimen Answer
Drawing 5—Draw the polygon of forces shown to a scale of 10 mm represents 1 newton. *R*—the *resultant*—is the line showing the direction and magnitude of a single force which could replace forces *A*, *B* and *C*.
Note—The direction of the resultant is *opposite* to the flow of the other three forces in the polygon.

Exercises
Exercises 1 to 4 can be worked to a scale of 10 mm represents 1 N.
1. By drawing a polygon of forces find the magnitude of forces *C* and *D* in order that O will remain in equilibrium—*Drawing 6*.
2. *Drawing 7*—Find the magnitude of forces *C* and *D* in order that O will remain in equilibrium.
3. *Drawing 8*—Draw a polygon of forces for the system shown. From your polygon calculate the magnitude of forces *D* and *E*.
4. *Drawing 9*—Find, by drawing, the magnitude of the forces *D* and *E*.
5. *Drawing 10*—Three forces *A*, *B* and *C* of 10, 15 and 20 N respectively are acting upon O. Determine the direction and magnitude of the resultant of the three forces. Work to a scale of 5 mm represents 1 N.

6. *Drawing 11*—Four forces *A, B, C* and *D* of 10, 15, 20 and 20 N respectively are attached to the object 0. In which direction will the object O move?

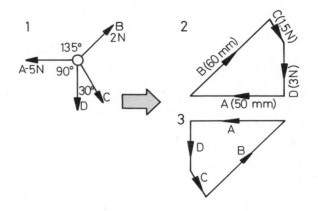

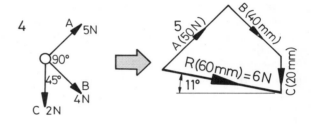

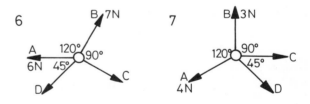

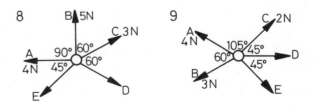

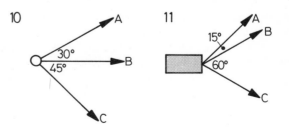

Drawing 1 shows a pin-jointed frame which is simply firm. Forces of 50 newtons and 200 newtons act on the two upper pin-joints. To hold the frame stationary (in equilibrium), the frame is placed on two supports at the lower pin-joints. Make rod DF 5 m long and rods AE and CF 2.6 m long.

To draw a stress diagram for the frame.

1. Draw the frame—scale 25 mm represents 1 m. Add letters—Bow's notation—in the spaces between bars.

2. Draw the triangle of forces for the top left hand joint—*Drawing 2*. Scale, 1 mm represents 2 N.

By measurement on this triangle it can be seen that a force of 58 N acts along bar AE *towards* the pin-joint. There is also a force of 30 N acting along bar *towards* the pin-joint.

3. Now that two of the forces acting on the top right hand pin-joint are known a polygon of forces can be drawn for that joint—*Drawing 3*.

Note—Force *BE* must act *towards* the right hand joint to counteract the force along *BE* acting towards the left hand joint. In other words there are forces in rod BE acting so as to resist the compression of rod BE caused by the action of the 50 N and the 200 N forces.

4. Now that two of the forces acting at the bottom left hand pin-joint are known, a polygon of forces for that joint can be constructed—*Drawing 4*.

5. Finally the triangle of forces for the bottom right hand pin-joint can be drawn—*Drawing 5*.

6. It is usual practice to draw a stress diagram incorporating all the triangles and polygons of forces for the frame—*Drawing 6*.

7. From this stress diagram all the forces acting on and in the frame can be calculated.

Note—It is usual practice to show members of a frame in *compression* by thick lines and members in *tension* by thin lines.

Compression and tension

Each bar in a *simply firm pin-jointed* frame is acted upon by equal and opposite forces at each pin joint. These forces in the bars result from the application of external loads applied to the frame and the reaction forces needed to hold the frame stationary (in equilibrium) at its supports.

Take two bars in the example given—bars BE and DF. In bar BE, equal and opposite forces are acting *towards* the pin joints holding BE. It follows therefore that the pin joints must be exerting equal forces on the bar in an opposite direction to the forces which the bar is exerting at the joint. Therefore bar BE is in *compression* under the action of the forces at the pin joint.

Similarly in bar DF equal and opposite forces are acting *inwards* away from the pin joints holding bar DF. Thus the pin joints must be exerting equal forces opposite to those in the bar. The bar DF is therefore under tension at the pin joints—*Diagram 8*.

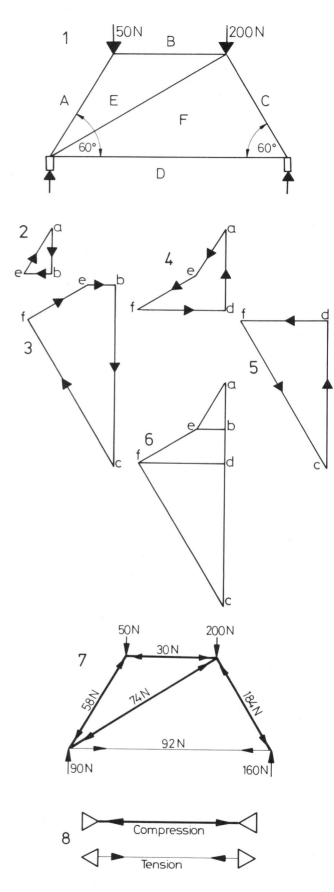

Exercises

1. By drawing a stress diagram for the given pin-jointed framework, find the forces acting in the bars of the frame and whether they are compression or tension forces. Also determine the magnitude of the reactions R and S at the supports for the frame.

2. Construct a triangle of forces to find the stresses in the two rods AB and BC.

3. Three horizontal forces are required to move the mass M along a horizontal plane. Find, and state, the magnitude and direction of a single horizontal force which could replace the three given forces.

4. A look-out cabin is supported by several pin jointed vertical frames of the shape ABCD. The cabin imposes a total load of 5000 N on each frame distributed in the ratio 3:2 acting vertically downwards at the joints A and B respectively. Draw an accurate stress diagram for one of the frames. From your diagram, determine:
   (a) the magnitude of the stresses in each member of the frame;
   (b) which members are in tension and which are in compression.

5. A radar aerial of parabolic section is mounted on a series of pin-jointed frames (ABCD) each made of light rods. Rods RA and SB connect the aerial to the frames and exert forces vertically downwards at A and B in the plane of each frame. Details are given in the drawing. **Construct a stress diagram for ABCD. Force RA = 200 N and force SB = 300 N. From your diagram determine:**
   (a) which members are in tension and which are in compression;
   (b) the forces acting along each rod;
   (c) the reactions at the pillars at C and D.

6. ABCD is a section through a roof which is to be covered with glass and roofing sheet. The roof is to be supported on a series of pin jointed frames resting on the walls at A and D.
   (i) Make a drawing of the given section to a scale of 1 in 100 with loads uniformly distributed.
   (ii) Add just sufficient members to make the frame rigid.
   (iii) Assume that each frame carries weights of glass and asbestos as given below and that the loading of the roof acts only vertically downwards. Ignore weight of frame members.
   Glass between A and B—600 kg; asbestos between B and C—350 kg; glass between C and D—400 kg.
   Show loads at each pin joint and at A and D which are proportional to the total load on the roof.
   (iv) Construct a stress diagram for the frame based upon the proportional loading.
   (v) From the stress diagram find:
   (a) the stresses in each member;
   (b) which members are in compression and which are in tension.

7. A hoist made from five light rods jointed freely at A, B, C and D is shown. Ignoring the weight of the rods, determine by graphical methods:
   (i) the forces acting along each of the five rods;

(ii) which of the rods are in tension and which are in compression;
(iii) the minimum force F required to prevent the hoist falling over under the action of the 500 kg weight.

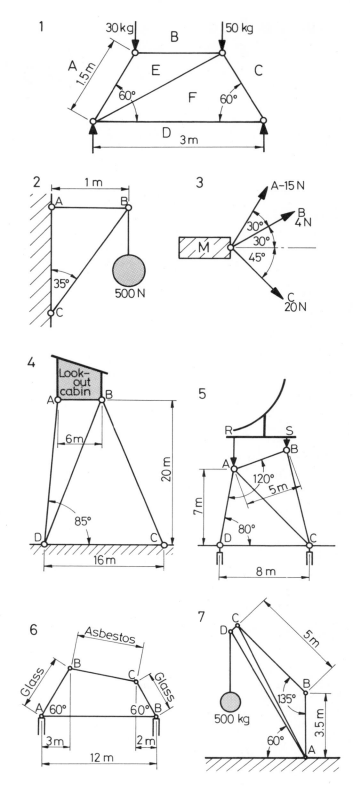

57

Link mechanisms

Many machines depend upon link mechanisms of various types in order to function. A few such link mechanisms are shown here. The student would be advised to examine machines to find other examples of link mechanisms.

Construct the apparatus shown in Drawing 1. Part A is a circular disc made from wood, or from metal, or from plastics or could be a pulley wheel. Part B—the link—can be made from Meccano strip, strip metal, strip wood or from plastics. Part C and the other parts are best made from wood. The base could be plywood. Part C should be capable of free movement along its slide. A plan of the mechanism, including suitable dimensions in millimetres is given in Drawing 2. This apparatus is a simple model of a common system to convert the circular rotation of part A into the linear motion of part C. You should be able to name many examples where this type of linkage is employed in machine mechanisms.

Place a piece of paper on the baseboard under the holes drilled along the link strip. Turn the disc A and, as the link moves, place a pencil point through each of the holes in turn so that the outlines of the movements of the holes are drawn. The three shapes you have drawn are known as the loci of the holes—singular, locus.

Pantograph

Using Meccano strip or strips of wood make up the linkage system of Drawing 3. Part DP should be about 300 mm long. It is important that the four strips form a parallelogram otherwise the mechanism will not work. Place a pencil point in a hole at A and move the end B so that the hole at B follows the outline of a shape. The pencil at A should make an exact replica of the shape but to a smaller scale. Repeat by placing the pencil at B and move the pantograph so that the hole at A follows a pattern outline. In this case the pencil should trace a replica outline, but to a larger scale at B.

Drawing 4 shows why the pantograph works in drawing to an enlarged or to a reduced scale. The triangles PBD and ABC are similar so all movements of B will be proportional to movements of A in the ratio PB:PA.

Drawings 5 and 6 show two forms of mechanical linkage such as may be found in some machines.
Drawing 5—The rotary movement of the arm PL moving through the arc of a circle is translated into linear movement of the rod CR through a bush.
Drawing 6—The rotary movement of the pivot C about its centre O is translated into straight line movement of the slider S in its slide.

Exercise

Design and make a model of the linkage system shown by Drawing 6.

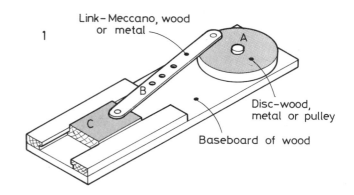

1 Link—Meccano, wood or metal — A — Disc—wood, metal or pulley — Baseboard of wood — B — C

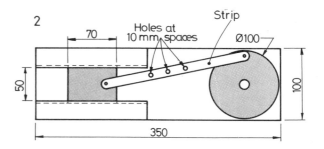

2 Holes at 10 mm spaces — Strip — 70 — Ø100 — 50 — 100 — 350

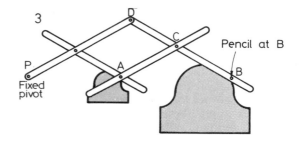

3 D — C — Pencil at B — P — A — B — Fixed pivot

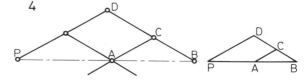

4 D — C — P — A — B — D — C — P — A — B

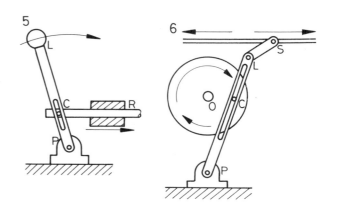

5 L — C — R — P — 6 — S — L — C — O — P

58

The five photographs on this page show examples of parts of technology projects carried out in schools.

Two girls working at a project to produce a model from strips of wood which will demonstrate instantaneous centres of rotation in mechanical linkages.

Part of a water-wheel project for the production of power from natural sources of energy—in this example—falling water.

Part of a light following trolley. The photograph shows the platform on which an electronics circuit is to be mounted.

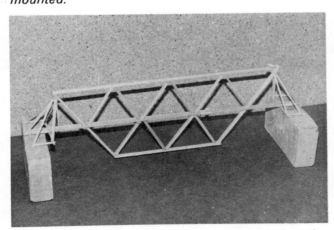

One of a series of girder support models designed as part of a project on structures.

Part of a project based on moving machines—power sources. In this example a model turbine wheel is under test.

Mechanisms

Pulleys

In the study of mechanisms a pulley is a *machine*. When making calculations relating to pulley systems, the following ratios and expressions are important: Mechanical Advantage (MA); Velocity Ratio (VR); Work; Efficiency.

Mechanical Advantage

$$\text{Mechanical Advantage—MA} = \frac{\text{load}}{\text{effort}}$$

In the examples given in the following pages, *grams* will be the units used in the calculation.

Velocity Ratio

$$\text{Velocity Ratio—VR} = \frac{\text{velocity of effort}}{\text{velocity of load}}$$

In the examples given here, the velocities of effort and load are taken by comparing the *distances* through which effort and load have moved.

Work

Work (W) is measured in *joules*. A joule is the work done by 1 newton of force moving through 1 metre of distance. The unit of work is the newton metre (Nm). *Note*—see page 49. 1 kgf = 9.8 newtons.

Efficiency

In the absence of friction MA should = VR. In any machine however, frictional and other losses will be present. Thus, in practice, MA is *always* less than VR. Because of losses such as by friction, *no* machine can be 100% efficient.

In the examples of pulleys given here, efficiency can be measured by:

$$\text{Efficiency} = \frac{\text{work done on the load}}{\text{work done by the effort}} \times 100\%$$

Pulley systems

The photograph above shows a stand and bracket on which pulley systems can be mounted. The stand can be constructed from a variety of materials:
1. A wood frame with supports made of Meccano strips.
2. A wood frame with wooden strips for the support. The wood strips can be screwed or nailed to the frame.
3. A wooden frame with metal strips screwed or bolted to the frame.
4. An all metal strip construction assembled with bolts.

Simple single pulley

Set up the simple single pulley system as shown.

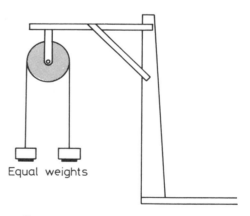

Equal weights

Single pulley system

1. Place equal weights on each side. Thus effort equals load. Make a note of what happens.
2. Add just sufficient weight on one side to start the load rising. Note the weight added. Why is this weight needed to raise the load?
3. Place twice the weight each side; then three times the weight; then four times; and so on. In each case add just sufficient weight on one side to start the load moving. Is the extra weight the same in each case?

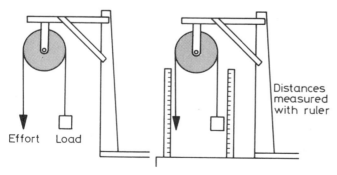

Single pulley system   *Measuring travel of effort and load*

Effort   Load

Distances measured with ruler

4. Work out the MA for each case. $MA = \dfrac{\text{load}}{\text{effort}}$.

5. Why is this less than 1 in each case?
6. How can the MA be improved?
7. If there is no MA in using the simple single pulley, why is it used?
8. Set up the single pulley shown, with a load on one side and a spring balance attached to the other side. At each of the angles indicated below note the readings shown by the spring balance in each case. The angles may be measured with the aid of set squares.
(a) To hold the load stationary
(b) To just lift the load
(i) vertical;   (ii) 30° to vertical;   (iii) 45° to vertical;   (iv) 60° to vertical;   (v) horizontal.

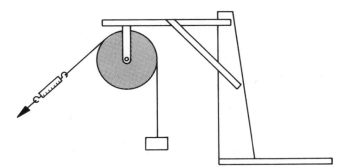

Simple single pulley

Simple pulley system with spring balances
Set up the single pulley system with two spring balances as shown, on the same stand as used for the simple single pulley.

Hang a weight from the pulley. Note the readings on each balance.

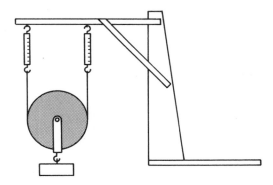

Single pulley system with two spring balances

Simple two pulley system
1. Set up two pulleys as shown with two spring balances in the system. Hold end A. Note reading on each spring balance. Hang a weight on the lower pulley. Note the reading on the spring balances.

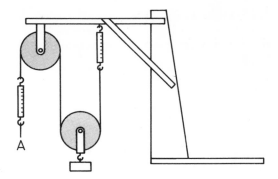

Two pulley system

Compare these readings with the first readings.
2. Pull on cord A to just raise the load. Note the readings on the spring balances.
3. Repeat 2 with other weights. Note all readings.
4. Work out the MA for each load. Does the MA vary with each load?

Alternative arrangement
Set up the alternative arrangement for a two pulley system as shown. Repeat the experiments as for the two pulley system with balances. Note all readings on the spring balance. Work out the MA for each load.

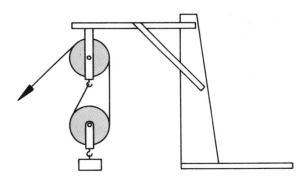

Alternative two pulley system

Work done by two pulley system
1. Remove the spring balance from the alternative two pulley system shown. Set up rules to measure the distances moved vertically by the load and vertically by the effort in the system.
2. Calculate the work done by the effort and the load. Remember
  (a) 1 kgf = 9.8 newtons
  (b) Work (W) = Nm joules where N is the force in newtons and m the distance travelled in metres.
3. Determine the efficiency of the two pulley system using the formula

$$\text{Efficiency} = \frac{\text{Work done on load}}{\text{Work done by effort}} \times 100\%$$

The spindle of the small electric motor shown in the photograph on the right revolves at 6000 revolutions per minute (rev/min) when connected to a 4.5 V battery.

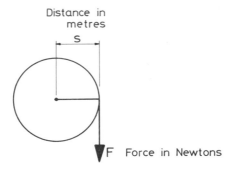

Distance in metres
s

F Force in Newtons

Torque = Fs

At this speed, this motor develops a *torque* (*T*) of 205 gram centimetres (gcm). In SI units the torque would be rated as 0.02 newton metres.

In this instance torque can be said to be a measure of the product of the force (*F*) in newtons (N) which can be exerted by the spindle revolving at 6000 rev/min., and the perpendicular distance (*s*) between the spindle axis and the line of the force. See drawing.

∴ for this electric motor at 6000 rev/min.

$$T = Fs = 0.02 \text{ Nm}.$$

If a pulley of diameter 20 mm is mounted on the motor spindle what is the value in newtons of the force exerted by the rim of the pulley?

$$T = 0.02 \text{ Nm}$$

∴ $0.02 = F \times 0.01$ (radius of pulley in metres)

$$\therefore F = \frac{0.02}{0.01} = 2$$

∴ The force will be 2 newtons. This is equivalent to a force of approximately 200 grams.

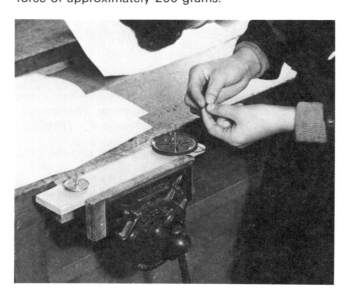

1. Set up two pulleys on a board, connected by a rubber band. Select pulleys of diameters where one is twice the diameter of the other. Rotate the large pulley through one complete revolution. Through how many complete revolutions does the smaller pulley turn?

2. Change the pulleys to others of different diameters. Rotate the largest of each pair in each case and check the revolutions turned by the other.

3. From the results of 1 and 2 state the rule which relates pulley speeds to pulley diameters in such pulley systems.

4. Set up three pulleys as shown in a diagram. By rotating the large single pulley determine the speed of the small single pulley. Does the result agree with the rule stated in 3?

5. Try other combinations of different sizes of pulleys set up as in the diagram. Check whether or not the rule applies to all these.

6. Set up a pulley system connected to an electric motor spindle in which the last pulley in the system rotates at one tenth of the speed of the motor.

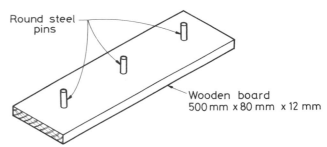

Round steel pins

Wooden board
500 mm x 80 mm x 12 mm

Board for pulley systems

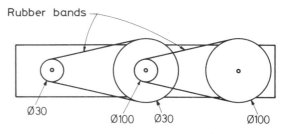

Rubber bands

Ø30   Ø100 Ø30   Ø100

Four pulley system

Model electric crane

To complete the project described on this page you will require the following:

1. A small electric motor capable of being driven by a battery.
2. A battery, a switch and connection leads or cables.
3. Materials from which a framework can be constructed to carry the motor, battery and pulleys. These materials could be:
   (a) Meccano strips, bolts and nuts, or
   (b) wooden strips glued and pinned, or
   (c) metal strips with holes, bolted together.
4. A number of pulleys and pulley belts.
5. A ruler and a watch capable of recording seconds.

Circuit

Connect the motor to the battery via the circuit shown in Drawing 1. Switch ON. In which direction does the motor spindle revolve?

Reverse the polarity of the battery in the circuit as in Drawing 2. Switch ON. What has happened?

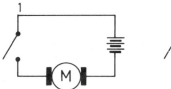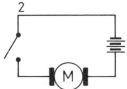

The torque of the motor

Assume the motor has the same rating as that described on page 62 opposite. Its torque is shown as 205 gram centimetres at 6000 rev/min. The motor you will be using should have its torque and rotation speed shown either on the motor, or on information packed with the motor.

Work out the torque (T) of the motor you are using in newton metres (N m). See page 62.

Another way of demonstrating the torque of the motor described on page 62 is to assume that if a pulley is attached to the spindle of the motor and a string wound around the pulley, the motor when running should be capable of lifting 205 grams using a pulley of 1 cm radius OR of lifting 1 gram using a pulley of 205 cm radius. The speed at which weights can be lifted can be varied by different pulley systems attached to the motor.

Note—The stated torque can never be fully achieved because of losses due to friction within the system of pulleys and spindles.

The project

Using the materials and the motor, design a crane system capable of lifting 10 grams through 200 cm in 1 minute. Write up a report of the project. Your report should include:
Sketches of the framework you have designed.
Sketches of the electric circuit.
Calculations of the pulley sizes.
A record of the weights, speeds and torque achieved.

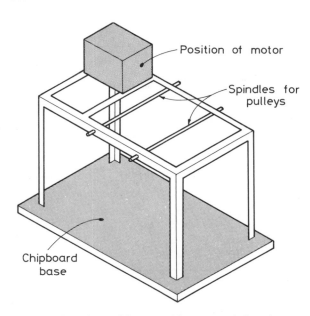

A suggestion for a Meccano framework for the crane

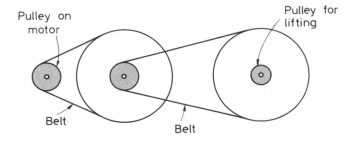

Suggested pulley systems

Levers

A *lever* is another type of *machine*. Drawings 1, 2 and 3 show, in diagrammatic form, three types of lever. In each, a load (*L*) can be moved by a force (*F*) with the lever acting around a point known as a *fulcrum*.

Exercise

Write down examples of three practical applications of each of the three types of lever shown in Drawings 1, 2 and 3, e.g. 1—prising a lid from a food tin; 2—lifting a wheelbarrow; 3—lifting a spadeful of earth.

Moments

The rotation of a lever around its fulcrum will be either in a clockwise or in an anti-clockwise direction. In the examples shown, the forces *F* applied to the levers are said to be producing *moments* in either a clockwise or in an anti-clockwise direction around the fulcrum. Moments in levers are measured by multiplying the force applied by the distance at which the force is working from the fulcrum or by multiplying the load by its distance from the fulcrum. The unit of moment is the newton metre or N m, or the newton millimetre (N mm). The two moments in lever systems—force × distance to fulcrum and load × distance to fulcrum—are equal.

Lever systems

1. Make the apparatus shown in Drawing 4. Move the 1 kg weight and the spring balance so that:
(a) $A = 40$ cm; $A1 = 40$ cm.
(b) $A = 30$ cm; $A1 = 40$ cm.
(c) $A = 20$ cm; $A1 = 40$ cm.
(d) $A = 10$ cm; $A1 = 40$ cm.
Multiply the distances *A*, in each case, by the 1 kg to give the anti-clockwise moment about the fulcrum; multiply the force measured by the spring balance in each case by the 40 cm. Work in N mm, remembering that 1 kgf = 9.8 N (or approximately 10 N). Make notes of your results. State your conclusions.
2. Make the apparatus shown in Drawing 5. Write down the anti-clockwise and clockwise moments of the 1 kg weight and the force F as measured on the spring balance when the distances *B* are:
(a) 60 cm   (b) 50 cm   (c) 30 cm   (d) 20 cm.
Make notes of your results. State your conclusions.
3. Repeat the two above projects with the pieces of apparatus shown in Drawings 6, 7, 8 and 9. Make notes of the clockwise and anti-clockwise moments in each example when the distances *C, D, E* and *F* are:
(a) $C = 10$ cm   (b) $C = 30$ cm   (c) $C = 60$ cm
(d) $C = 70$ cm   (e) $D = 10$ cm   (f) $D = 20$ cm
(g) $D = 60$ cm   (h) $D = 80$ cm   (i) $E = 80$ cm
(j) $E = 60$ cm   (k) $E = 30$ cm   (l) $E = 10$ cm
(m) $G = 80$ cm   (n) $G = 50$ cm   (o) $G = 20$ cm
(p) $G = 10$ cm

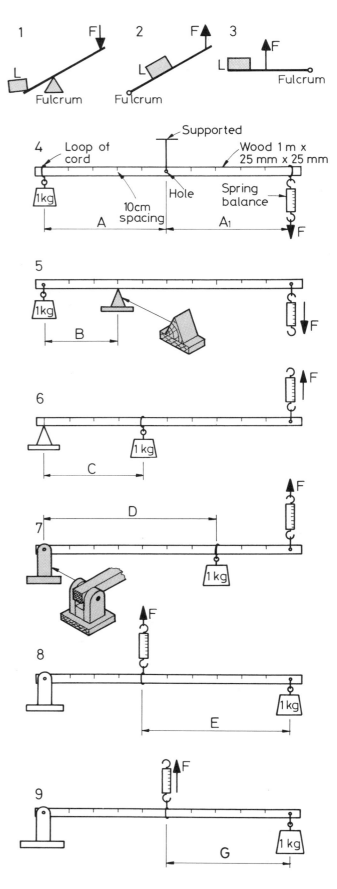

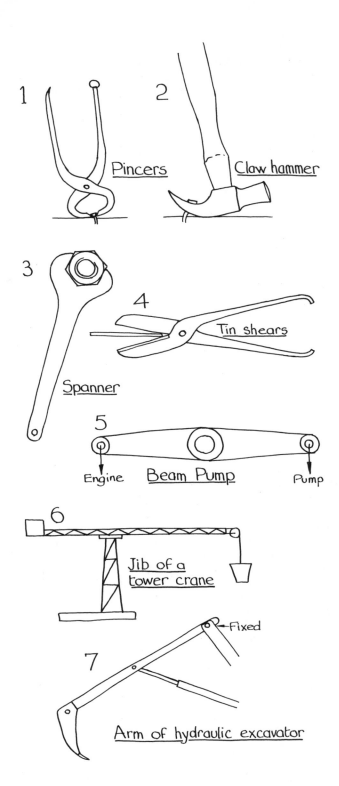

1 Pincers

2 Claw hammer

3 Spanner

4 Tin shears

5 Beam Pump — Engine — Pump

6 Jib of a tower crane

7 Fixed

Arm of hydraulic excavator

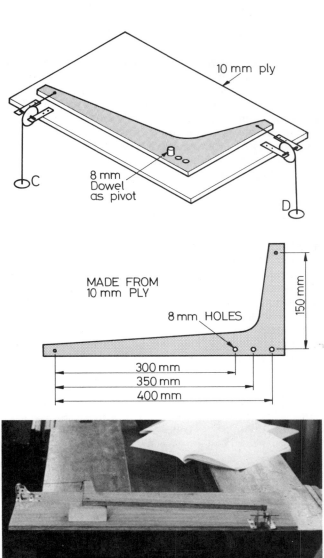

10 mm ply

C

8 mm Dowel as pivot

D

MADE FROM 10 mm PLY

8 mm HOLES

150 mm

300 mm

350 mm

400 mm

Bell crank lever

A common lever device employed in machinery is the bell crank lever. A drawing of an experimental bell crank lever which you could make yourself is shown, together with a photograph showing the experiment being worked.

Three holes A, B and C, are shown in the drawing giving dimensions for the lever. Using each hole in turn as a pivot point (fulcrum) and with a 1 kg weight hung at C, find, both by experiment and by calculation, the weight to be hung at D to balance the weight at C. Calculate the moments acting at each of the holes A, B and C.

Exercises

A number of tools and other devices are shown in freehand outline sketches 1 to 7. In each example state with the aid of sketches: (a) where the fulcrum is; (b) at which point a force must be applied; (c) the position of the load on which the lever action of the tool or device is working.

Moments

1. Construct the apparatus shown in Drawing 1. The beam consists of a strip of wood of a section about 35 mm by 25 mm marked at regular intervals of 10 cm along its length. Note that 10 cm is 100 mm or 0.1 m.
2. Place a balance directly under 1 on the beam. Each time a weight is to be hung on the scale, adjust the balance to zero after placing the weight carrier in a new position. This will prevent errors arising in your calculations due to the weights of the beam and the weights carrier.
3. Hang a 1 kilogram (kg) weight on the carrier at division 2 on the beam. Note the reading on the balance.
4. Move the 1 kg weight to hang at 4. Note balance reading. Repeat at division 5.
5. Move the balance to division 2 and hang the weight at divisions 4, 6 and 10. Note all readings.
6. Calculate the moments clockwise and anti-clockwise around the pivot P from each of your noted readings.

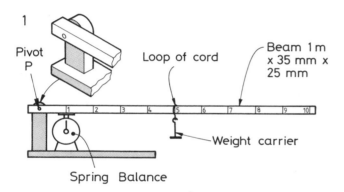

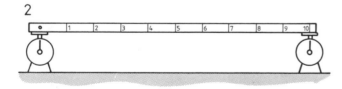

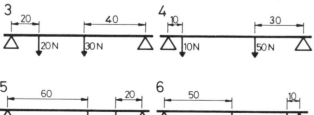

Dimensions in mm

△ Balance   ↓ Force in Newtons

Worked example

Take the example when the balance is under division 1 and the 1 kg is hung at division 4.

Take 1 kilogram force as equal to 10 newtons.

Moments clockwise around P

The weight of 1 kg is acting downwards as a force of 10 N, 400 mm from P. Its moment is therefore 10 N × 400 mm = 4000 N mm (newton millimetres) or 4 N m (newton metres).

Moments anti-clockwise around P

The reading on the balance should be 4 kg. This represents a force (in the spring of the balance) acting vertically upwards, of 40 N, which is acting at a distance of 100 mm from P. Its moment is 40 N × 100 mm = 4000 N mm or 4 N m.

Thus the two moments around P are equal and opposite.

Moments acting on a beam

Using the same beam, set it on two balances as in Drawing 2. Hang weight carriers on the beam at divisions 4 and 8 and adjust both balances to zero (see Drawing 3). Hang a weight of 2 kg on the carrier at division 4 and a 1 kg weight on the carrier at division 8. Note the readings on the balances. That at division 1 should read 1.6 kg. That at division 2 should read 1.4 kg. Note that the two readings add up to 3 kg—the total of the two weights on the carriers.

Calculation of moments

Clockwise around division 1

20 N (the 2 kg weight) × 400 mm + 10 N (the 1 kg weight) × 800 mm = 8000 + 8000 = 16000 N mm or 16 N m.

Anti-clockwise around division 1

The anti-clockwise moment is the force (F) exerted vertically upwards by the spring of the balance at division 10 multiplied by its distance from division 1. This can be written as 1000 F N mm, where F is the force shown at balance, since clockwise and anti-clockwise moments around a point should be equal.

1000 F = 16000 N mm

∴ F = 16 N (the reading of 1.6 kg on the balance).

Clockwise around division 10

10 N × 200 mm + 20 N × 600 mm = 2000 + 12000
$$= 14000 \text{ N mm}$$
$$= 14 \text{ N m}.$$

Anti-clockwise around division 10

This is 1000 F_1 where

F_1 = Force shown at balance.

Thus 1000 F_1 = 14000 N mm

∴ F_1 = 14 N (the reading of 1.4 kg on the balance).

Exercises

Calculate the readings in each case which you would expect to find on the balances if the beam and its weights were set up as shown in Drawings 3, 4, 5 and 6.

A moments experiment

around a piece of wooden dowel of 25 mm diameter. State the pitch of each of the three helices formed on the dowels.

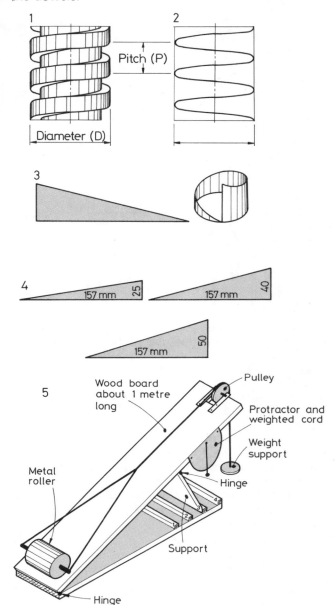

The screw thread of a Rockwell hardness testing machine

The screw

A *screw* is another type of *machine*. In practice there are a large number and variety of types of screw threads, of which the *square thread* shown in Drawing 1 is an example. Such a screw is made up of a number of helices, one of which is shown in Drawing 2. Note the two terms—diameter (*D*) and pitch (*P*). The diameter of a screw can be taken as its external diameter. The pitch of a screw is the distance, measured along its axis, taken by one complete turn of one of the helices forming the screw.

If one of the helices (singular is helix) could be 'unwrapped' it would be seen that screw threads are similar in their action to inclined planes—*Drawing 3*.

Exercise

Mark out on a piece of paper and cut to shape the three triangles shown in Drawing 4. Wrap each of them

Experimental apparatus

Make up the inclined plane apparatus shown in Drawing 5. Weigh the metal roller. Note its weight. With the inclined plane support first at 1, then at 2, then at 3 and so on, find the weights that have to be hung on the pulley to cause the roller to just roll up the plane. Make a note of the results in each case. What conclusions do you draw from these results?

Note—Because of frictional losses acting on the roller by the surface of the plane and the friction of rotation of the pulley spindle, your results will not be perfect. They should however be sufficiently accurate for you to draw reasonable conclusions.

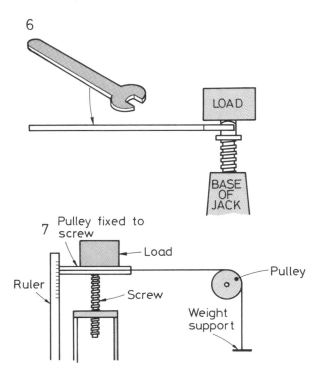

6

7 Pulley fixed to screw — Load
Ruler — Screw — Pulley
Weight support

There are numerous applications of the screw acting as a machine. For example the action of a woodwork vice screw or a metalwork vice screw, holds materials firmly while you work on them. One example of the action of a screw is in the lifting properties of a screw jack (Drawing 6). In order to lift a load, a force is applied at the end of a spanner operating the jack. This force applies a torque (T) to the screw via the spanner. The work done by the force is measured by the distance through which the force travels. In each complete revolution of the spanner this distance = length of spanner (L) $\times 2\pi$. Therefore in one revolution of the spanner, the work done in newton metres = force $\times L \times 2\pi$. If *frictional losses are ignored* the work performed by the screw is:

$$\text{load} \times \text{pitch},$$

and since the work done on the screw = the work performed by the screw, force $\times L \times 2\pi$ = load $\times$ pitch.

Example 1
What force in newtons must be applied to a spanner of 200 mm length to lift a load 50 kg using a screw jack, whose screw pitch is 5 mm?
Answer Let 1 kgf = 10 newtons.

$$F \times 200 \times 2 \times 3.14 = 500 \times 5$$
$$\therefore F = \frac{500 \times 5}{200 \times 2 \times 3.14} = \frac{25}{12.56} = \text{approximately 2 N}$$

Example 2
If a force of 10 newtons is applied at the end of a spanner of 250 mm length working a screw jack with a screw of 4 mm pitch, what weight will the jack lift?
Answer Let 1 kgf = 10 newtons.

$$10 \times 250 \times 2 \times 3.14 = L \times 4$$
$$\therefore L = \frac{10 \times 250 \times 2 \times 3.14}{4} = 3925 \text{ N}$$
$$\therefore \text{Load is 392 kg approximately.}$$

Exercises
In these exercises let 1 kgf = 10 N; P = pitch of screw; L = load; F = force in newtons.
1. A nut of 3 mm pitch is being tightened on to a screw thread with a spanner of 200 mm length. If a force of 20 newtons is applied on the end of the spanner, what force is acting between the nut and its screw?
2. If a load of 0.5 kg is being lifted with a jack of 10 mm screw pitch with a working arm of 0.5 m long, what force is necessary on the arm to lift the load?
3. If a load of 0.2 kg is to be lifted by a screw of 5 mm pitch with a force of 10 newtons, how long a spanner will be required?

Mechanical advantage of a screw
If a force of 2 newtons acting on a screw via a spanner will lift a load of 10 kg, then its mechanical advantage will be the load in newtons divided by the force in newtons,
$$\therefore \text{MA} = 10 \times 10 \div 2 = 50$$

Velocity ratio of a screw
If a screw with a pitch of 10 mm is moved through 1 pitch by a spanner of length 200 mm then its velocity ratio (VR) is the distance moved by the force applied to the spanner divided by the pitch. In this example, the distance moved by the force applied to the spanner is $200 \times 2 \times \pi$ mm, the distance moved by the screw is 10 mm, $\therefore$ VR = $200 \times 2 \times \pi \div 10 = 125.6$.

Exercises
Calculate the mechanical advantage and the velocity ratio in each of the three exercises given earlier on this page.

Experiment
Design an experimental device based on the diagrammatic Drawing 7. Use the device to verify the conclusions drawn from your work on screws. Remember that frictional losses throughout such a device will cause your results to be slightly inaccurate.

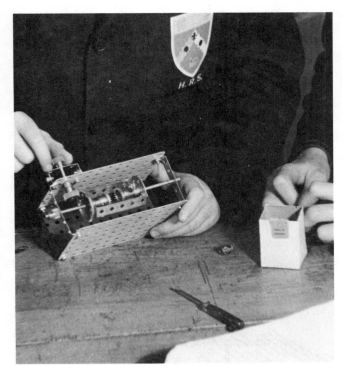

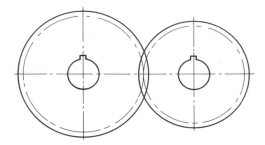

Conventional drawing of spur gears

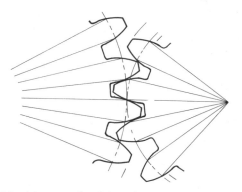

Meshing teeth of involute spur gears

Gears and gear trains

Gears are designed to transmit rotary movement from one part of a mechanism to another. There are several forms of gear teeth, the commonest form used in engineering being those whose teeth outlines are based on involute curves. A variety of types of gear, four of which are illustrated on this page and page 70 overleaf are available.

Spur gears—will transmit rotary movement in the same plane.

Bevel gears—will transmit rotary movement from one plane to another at an angle to the first.

Rack gears—will enable rotary movement to be converted to straight line movement, or vice versa.

Worm gears—will transmit rotary movement through an axial angle of 90°.

Spur gears

Spur gears will transmit rotary drive from one shaft parallel to another. A spur gear with a small number of teeth can be referred to as a *pinion* gear. Note the three simple spur gear arrangements shown.

Drawing 1—When two spur gears, each with the same number of teeth are meshed, the driven gear rotates at the *same* speed as the driving gear, but in the *opposite* direction.

Drawing 2—Spur gears are frequently used to change the speed of rotation of parallel shafts. The difference in rotational speeds of the shafts is directly proportional to the numbers of teeth on the gears. Thus a driving gear of 48 teeth, meshing with a driven gear of 16 teeth causes a 3-fold increase in the rotational speed of the driven gear—$48 \div 16 = 3$.

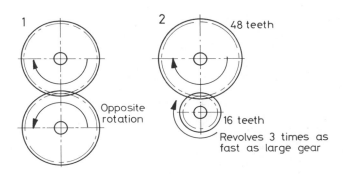

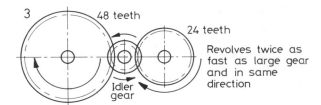

Simple spur gear arrangements

Note—(a) If the driving gear has 16 teeth and the driven gear 48 teeth then the driven gear rotates at $\frac{1}{3}$ of the driving gear.

(b) In Drawing 2 it can be seen that, although the speeds of the two shafts are different, they are rotating in the *opposite* direction to each other.

(c) If the driven and driving gears are to rotate in the *same* direction, an *idler* gear can be inserted between the two gears. The idler gear can have *any* number of teeth, but is usually a *pinion* gear.

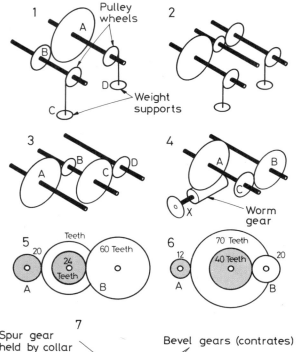

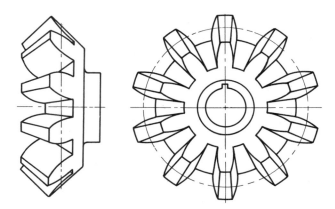

Front and end views of a ten teeth bevel gear

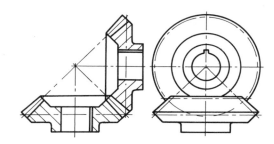

Conventional drawing of a bevel gear

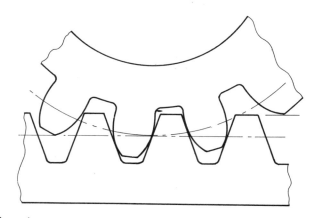

A rack gear

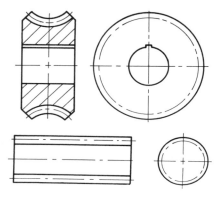

Conventional drawings of a worm wheel and a worm

Spur gearing

1. From the preceding page: the ratio of rotational speeds between driving and driven gears is directly proportional to the number of teeth on the gears.

2. If the speed of a shaft with a gear is reduced by meshing its gear with a larger gear, then the power drive of the shaft is increased in proportion to the reduction in the rotational speed. This is if frictional losses are ignored.

Exercises

1. Set up the gearing system shown diagrammatically in Drawing 1. Gear A should have four times as many teeth as gear B. Place a 20 gram weight at C. Place weights on D until C just begins to rise. Note the weight on D.

2. Repeat with gear A with twice as many teeth as gear B.

3. Repeat with gear A with three times as many teeth as gear B.

4. Set up the gearing system shown by Drawing 2 using the same gearing ratios as in Exercises 1, 2 and 3. Note the results. Gear X can be a small pinion gear.

5. Set up the gearing system shown by Drawing 3. Count the number of times gear D rotates when gear A is rotated through *one* revolution under the following circumstances:
(a) Gear A twice as many teeth as gear B; gear C twice as many teeth as gear D.
(b) Gear A twice as many teeth as gear B; gear C three times as many teeth as gear D.
(c) Gear A four times as many teeth as gear B; gear D three times as many teeth as gear C.

Note all your results and state your conclusions. Turn back to the pages describing pulleys and compare the results of working with gears with working with pulleys.

Note—The gear assemblies as shown in Drawing 3 are known as *compound* gear trains.

6. Set up the compound gear train shown in Drawing 4. Count the number of complete rotations of X needed for gear C to rotate one complete revolution when:
(a) A has twice as many teeth as B and C three times as many teeth as B.
(b) A has four times as many teeth as B and C has twice as many teeth as B.

7. If A is the driving gear in each example, calculate:
(a) The relative rotational speeds of B compared with A and
(b) the power increase or decrease of the shaft driven by gear B when compared with gear A, in the gear trains shown by end views in Drawings 5 and 6.

8. A gear differential system is shown diagrammatically in Drawing 7. It is also illustrated in a photograph on page 69. Attempt making up the gear differential system shown. What is the purpose of this system of gearing?

Cams

You will have noted that a *rack gear* system can be employed in a mechanism to convert rotary movement into straight line movement. Other devices for converting rotary movement along a line include radial cams. A variety of radial cams are illustrated on this page.

As a radial cam rotates, a follower resting against the profile of the cam is made to move up and down, or in and out as it follows the cam profile. The follower is usually held firmly against the cam profile under the action of gravity or under the action of a spring.

The drawings of cams shown here are all *sectional* drawings, the spindle or shaft of the cam being shown hatched.

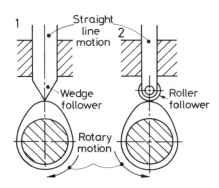

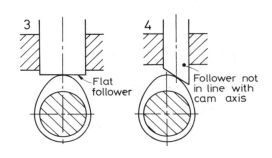

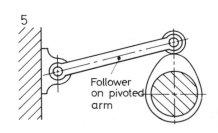

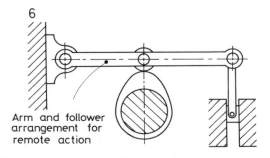

Some types of radial cam arrangements

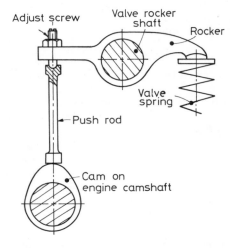

Overhead valve layout from a car engine

71

Electrics

Electrical circuits

An electric *current* may be regarded as a flow of ultra microscopic particles known as *electrons*. An electric *battery*, which is a collection of electric cells, may be regarded as producing electrons. When an electric battery is connected to a piece of electrical equipment, the electrons from the battery will flow through and operate the equipment. Electrons are *negatively* (− ve) charged and are produced at the negative terminal of a battery. The positive (+ ve) terminal of the battery attracts electrons. The *electron* flow is therefore from *negative* to *positive*. There is said to be a *potential difference* (PD) between the terminals of an electric battery.

Conductors are used to carry the electric current from the battery to the equipment being operated. For the experiments and projects described in the following pages, *insulated* copper wire cable is suitable. *Copper* is a good conductor of electricity. The *insulation* covering the cable prevents leakage of electricity from the equipment being operated. An *electric circuit* must be *complete* if the circuit is to operate properly. There *must* be a continuous line of conduction from the battery to the equipment and back to the battery. This *complete* circuit allows the electrons to flow from battery through the conductors through the equipment and back to the battery. Some of the electricity will operate the equipment. The energy of the electric current will have been converted into the energy produced from the equipment. This energy may take the form of heat, or light, or noise and so on.

The components in the electrical circuits shown in experiments in the pages which follow, can all be connected to each other with the aid of standard 4 mm stackable plugs and sockets. It is suggested that 0.2 mm copper insulated wire cable is a suitable conductor between the components. If red and black coloured plugs, sockets and wire are used, the positive and negative sides of circuits can be easily identified. Always try to keep the red plugs and sockets and cables on the positive side and the black plugs, sockets and cables on the negative side.

Rules

1. Circuits *must* be complete. *Check* before switching ON.
2. Reds to positive; blacks to negative.
3. *Check* circuits for correct bulb sizes etc. before switching ON.
4. When using *meters* check that the meter is of the correct rating for the circuit. If *shunts* are used with meters, check that the shunt is of the correct resistance for the circuit. If in doubt use meters which will carry a greater current or voltage than is expected.

Circuits symbols

The symbols shown on the right are those which are suitable for drawing circuits shown throughout this

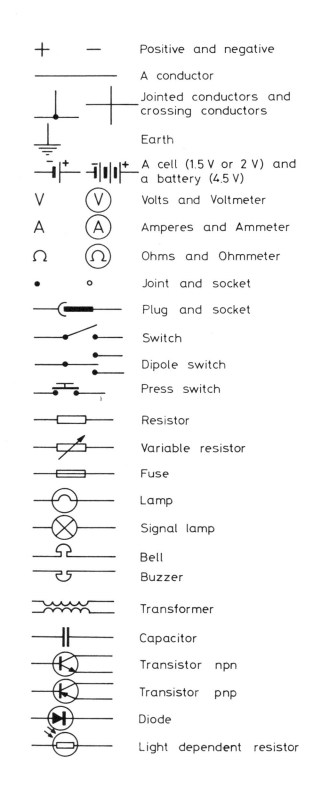

+ −	Positive and negative
	A conductor
	Jointed conductors and crossing conductors
	Earth
	A cell (1.5 V or 2 V) and a battery (4.5 V)
V Ⓥ	Volts and Voltmeter
A Ⓐ	Amperes and Ammeter
Ω Ⓞ	Ohms and Ohmmeter
• ○	Joint and socket
	Plug and socket
	Switch
	Dipole switch
	Press switch
	Resistor
	Variable resistor
	Fuse
	Lamp
	Signal lamp
	Bell
	Buzzer
	Transformer
	Capacitor
	Transistor npn
	Transistor pnp
	Diode
	Light dependent resistor

book. All these symbols have been taken from the British Standards publication BS 3939 (Graphical symbols for electrical power, telecommunications and electronics diagrams).

Methods of drawing electric circuit diagrams

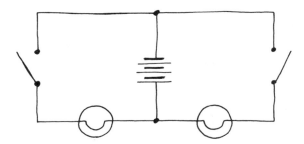

Freehand on plain paper

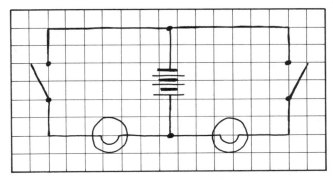

Freehand on square grid paper

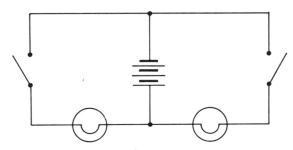

With the aid of instruments—ruler and compasses

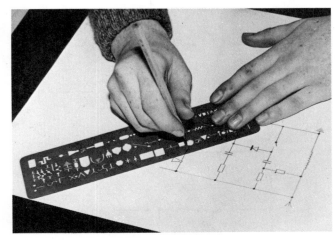

With the aid of a template

Equipment for circuit experiments

The illustrations below show a method of constructing modular size boxes for the components used in the simple electrical circuit experiments given on pages 74 to 81. Other types of 'modular' equipment are also illustrated on these pages.

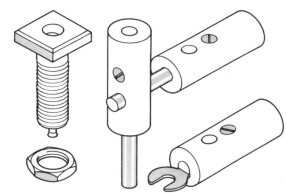

4 mm insulated socket *4 mm stackable plugs and spade terminal*

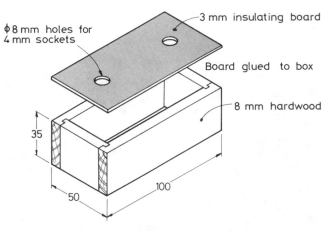

ϕ8 mm holes for 4 mm sockets

3 mm insulating board

Board glued to box

8 mm hardwood

35

50

100

Construction of modular component boxes

Switches and light bulbs

When an electric light bulb is lit by electricity, the energy of the electron flow is converted into energy in the form of heat. The heat energy causes the filament of the bulb to become white hot. The white heat of the filament gives off energy in the form of light.

The simple toggle switches which should be used to make up the circuits shown on this page *make* or *break* the continuity of the conductor in which they are placed.

Switch and bulb circuits

1. Connect a light bulb, a switch and a battery in the circuit shown. Check that the circuit is complete. Switch ON. What happens? Switch OFF. What happens? Suitable ratings are: battery—4.5 V; bulb—3.5 V, 0.3 A.

2. Add a second bulb. Switch ON. What happens? The two bulbs are said to be in *series*. Compare the light from each bulb with the light obtained from the single bulb in circuit 1. Switch OFF.

3. Replace the second bulb with a second switch, as shown. Switch one switch ON. What happens? Switch the second switch ON. What happens? Switch OFF either switch. What happens? The switches are in *series*.

Note—both Switch 1 *and* Switch 2 must be ON in order that the bulb lights.

4. Connect two bulbs, a switch and a battery as shown. Switch ON. What happens? The two bulbs are in *parallel*.

5. Connect two switches, a bulb and a battery as shown in the photograph and diagram. The switches are now in *parallel*. Switching on either Switch 1 or Switch 2 will light the bulb. If *both* switches are ON, what happens when one of them is switched OFF?

Exercises

Draw circuit diagrams for the following electrical circuits. Use symbols from the list shown on page 72. Set up the circuits you have drawn and test them to check whether or not your circuit diagrams are correct.

1. Three light bulbs, connected in parallel to a battery to be all switched ON or OFF from one switch.

2. Three switches connected in parallel to a battery to light a bulb by switching ON any one switch.

3. Two bulbs connected via a switch to a battery. One bulb lights when the other is OFF and vice versa.

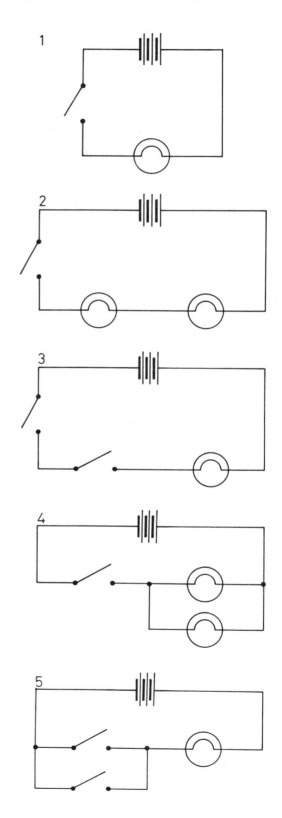

Burglar alarm

A night watchman in a building which has three entrances wishes to be able to identify which entrance has been used by any person entering the building. One method of doing this would be to fit an electrical switch on each door and connect the switch via a circuit to numbered light bulbs on an indicator board.

The toggle switches used to make up the circuits shown on page 74 opposite are not suitable for this purpose. Instead of toggle switches it is suggested that *micro switches* are used.

1. Set up a circuit containing a bulb, a battery and a micro switch. Try different connections on the micro switch and make a note of the results.

2. Design and make a simple device for attaching the micro switch to a door in such a manner that when the door is opened, the switch is in its ON position.

3. Write down in note and drawing form what has been carried out. Include a circuit diagram with correct symbols.

4. Design on paper, a circuit which would be suitable for the indicator board lights to operate from the three doors in the building.

5. A *reed switch* is operated when brought into the magnetic field of a magnet. Design a circuit and a device by which it could be fitted to a door, in which a reed switch could be used in place of a micro switch to operate the indicator board lights.

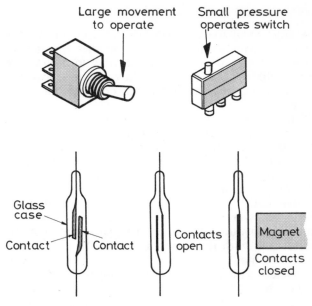

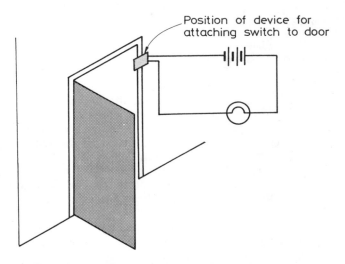

A reed switch

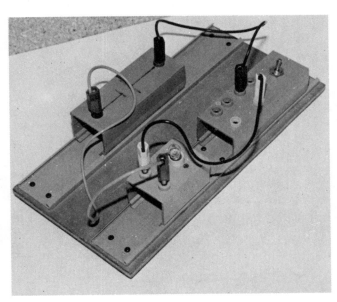

A 'Danum' kit in use. Model circuit for bedroom light (page 14)

In the example on the right a bell is being used in place of an indicator lamp

Measuring current

The amount of the electron flow in a circuit can be measured with an *ammeter.* An ammeter measures *current*—it measures the electron flow in a circuit in units called amperes (amps). In order to make this measurement, an ammeter must be connected in *series* in a circuit.

Three units of measurement are commonly used in the types of circuits described in this book—the ampere (A); the milliampere (mA) which is one thousandth of an ampere; the microampere (μA) which is a millionth of an ampere.

When using ammeters you must make sure they are not overloaded by measuring a greater current than the instrument was designed to measure. Ammeters are expensive instruments and can be easily burned out by overloading with too heavy a current. You must also make sure that the positive terminal of the ammeter is connected to the positive side of the electrical supply or the negative of the ammeter connected to the negative side of the supply.

For the work attempted on this page use a meter which reads up to 1 ampere in 100 mA. Note that 100 mA = 0.1 A.

1. Set up Circuit 1. Switch ON. Note the reading. The reading gives the quantity of current in mA. Switch OFF.
2. Change the ammeter to the other side of the battery supply. Make sure you connect the ammeter to the correct terminal—positive of ammeter to positive of battery or negative of ammeter to negative of battery. Switch ON. Note the reading. Compare the reading with that from circuit 1. Switch OFF.
3. Add a second bulb to the circuit so that the two bulbs are in series. Switch ON. Note the reading. Is the reading different from those from Circuits 1 and 2? Switch OFF.
4. Change the second bulb from Circuit 3 so that the two bulbs are in parallel. Switch ON. Note the reading. Compare this reading with those obtained on the other circuits. Switch OFF.
5. Place a second ammeter in the circuit to obtain circuit 5. Switch ON. Compare both readings with those made in other circuits.

Exercises

Draw the circuit diagrams for the following:
1. Two switches in series operating two bulbs in parallel with an ammeter in the circuit to measure the total current when *both* switches are ON.
2. A circuit containing a battery, switch and 4 bulbs in parallel with ammeter placed to measure the current flowing in each bulb.
3. A circuit with 2 bulbs in parallel which can be switched independently, with an ammeter measuring the current consumed by either one or both bulbs when switched ON.

Make up each of these three circuits and note the reading on the ammeters.

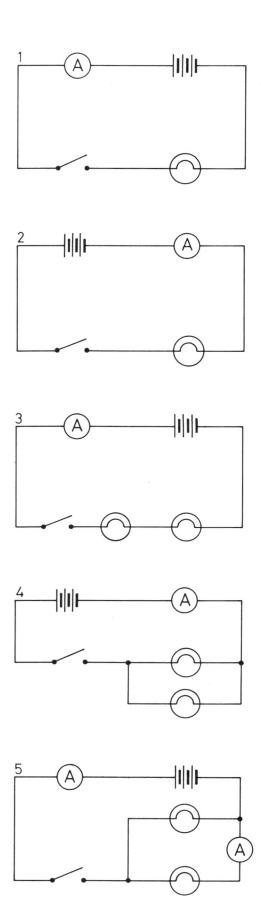

Measuring voltage

At the negative (−ve) terminal of a battery, it can be said that there is an accumulation of electrons, which will flow when the battery is connected into a closed circuit. When the battery is connected into a circuit, the accumulation of electrons at the negative terminal is being constantly replaced by chemical reactions within the battery. At the positive (+ve) terminal there is a shortage of electrons. The difference between the two terminals causes an electrical pressure between them. This electrical pressure is known as a *potential difference* (PD)—when a battery is connected into a circuit this electrical pressure (PD) causes electrons to flow through a circuit.

We already know that a *flow* of electrons is known as a *current*. The pressure causing the flow of electrons is measured in units called *volts* (V). To measure the potential difference in volts, a *voltmeter* is used. It should be noted that, to measure volts, a voltmeter must be connected in *parallel* in a circuit in order to accurately measure the potential difference in that part of the circuit being measured.

Voltmeters and ammeters are, basically, the same instruments. Some voltmeters can be adapted from ammeters by the addition of voltmeter resistors. Always make quite certain that either the correct voltmeter resistor is in use, or that the voltmeter you are using is designed to measure the maximum voltage you may expect in the circuit.

Points to remember
1. *Volts* measure pressure (PD).
2. *Amperes* measure current flow.
3. Voltmeters must be placed in *parallel* in a circuit.
4. Ammeters must be placed in *series* in a circuit.
For the work attempted on this page a voltmeter measuring up to 5 volts is required. The circuits described on this page all use a 4.5 volt battery.
1. Connect the voltmeter to the battery and note the reading.
2. Set up Circuit 2. Switch ON. Note the reading on the voltmeter.
3. Switch OFF. Change the voltmeter to a new position as shown. Switch ON. Note the reading. Why is it the same as in Circuit 2?
4. Switch OFF. Change the voltmeter back to the same position as for Circuit 2. Insert an ammeter (reading to 100 mA) in the circuit. Switch ON. Note readings on ammeter and voltmeter. Switch OFF.
5. Set up Circuit 5. Switch ON. What change of voltage reading has taken place?
6. Switch OFF. Change the voltmeter as in Circuit 6. Switch ON. Note the readings. Compare the readings on the voltmeter with those in Circuits 2 and 5.

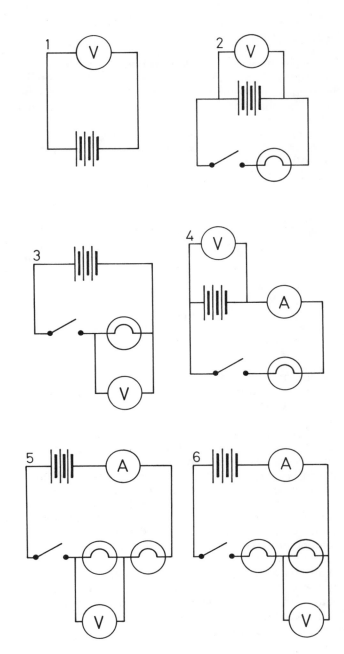

Exercises
Draw the following circuit diagrams:
1. Battery, switch, two bulbs in parallel; voltmeter testing voltage across *either* bulb.
2. Battery, switch, 2 bulbs in parallel. Voltmeter set up to measure voltage across *either* bulb. Two ammeters to measure current used in both the *whole* circuit and by *one* bulb.

Resistors

When a battery is connected to an electrical circuit and the circuit is switched ON, the quantity of current flowing in the circuit depends upon the *resistance* in the circuit to the flow of electrons and on the potential difference in the circuit.

Resistance, current and voltage are dependent upon each other. If R represents resistance, I the current and V the voltage then:

$V = IR$ and from this it follows that

$R = \dfrac{V}{I}$ and $I = \dfrac{V}{R}$.

A pictorial method of remembering these simple formulae is shown.

$$V = IR \qquad I = \frac{V}{R} \qquad R = \frac{V}{I}$$

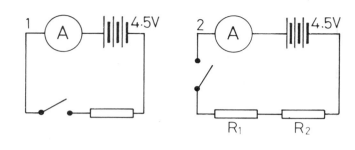

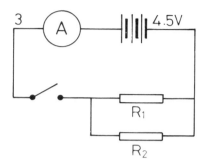

Resistance is measured in units called *ohms*. Ohms are represented by the symbol Ω. The equations given above are based on what is known as *Ohm's Law*, which states that:

If a circuit is connected to a 1 volt source of supply and 1 ampere of current is flowing through the circuit, then its resistance is 1 ohm.

The knowledge and use of Ohm's Law is of great importance in the understanding of electric circuits.

Use carbon resistors in the work shown on this page. The resistance, in ohms, of these resistors is shown by coloured bands on their surfaces. Charts showing these colours and their meanings are easily available.

Resistors in series

1. Set up Circuit 1. This should consist of a 4.5 volt battery, a $100\,\Omega$ resistor (brown, black, brown), a 100 mA ammeter and a switch. Switch ON. What is the reading on the ammeter? It should be 45 mA. Why?

Because: $I = \dfrac{V}{R} = \dfrac{4.5}{100} = \dfrac{45}{1000} = 45\,\text{mA}$.

Variations around 45 mA may occur because:
(a) the battery output may not be fully 4.5 V;
(b) carbon resistors have a tolerance of 10%. Thus a $100\,\Omega$ carbon resistor has a value between about $90\,\Omega$ and $110\,\Omega$.

2. Switch OFF. Add a second $100\,\Omega$ resistor into the circuit—Circuit 2. Switch ON. Note the ammeter reading. It should be approximately half of the previous reading. This is because the resistance in the circuits has been doubled. Thus, by Ohm's Law:

$I = \dfrac{V}{R} = \dfrac{4.5}{200} = \dfrac{2.25}{100} = 22.5\,\text{mA}$.

Note—when *resistors* are in *series* their total resistance is obtained by *adding* their values.

Resistors in parallel

3. Switch OFF. Change the second $100\,\Omega$ resistor so that it is in *parallel* with the first—Circuit 3 (see photograph). Switch ON. Note the reading on the ammeter. It should be double that obtained in Circuit 1—it should be 90 mA. From Ohm's Law: $R = \dfrac{V}{I} = \dfrac{4.5}{0.09} = \dfrac{450}{9}$

$= 50\,\Omega$

The formula for finding the value of resistors in parallel is: $\dfrac{1}{R} = \dfrac{1}{R_1} + \dfrac{1}{R_2}$.

Thus from Circuit 3: $\dfrac{1}{R} = \dfrac{1}{100} + \dfrac{1}{100} = \dfrac{2}{100} = \dfrac{1}{50}$

$\therefore R = 50\,\Omega$

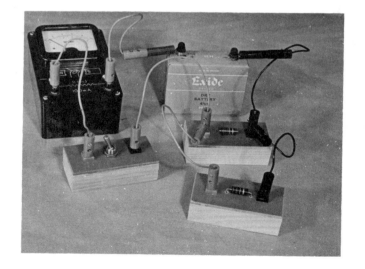

4. Repeat Circuit 1 using a 1000Ω resistor (brown, black, red). Switch ON. Note the reading. It should be $\frac{1}{10}$ of that of Circuit 1, because the resistance has been increased tenfold.

5. Repeat Circuit 2 using 1000Ω resistors. Switch ON. The reading should be $\frac{1}{2}$ that of Circuit 4.

6. Repeat Circuit 3 using 1000Ω resistors. Switch ON. The reading should be approximately 9 mA.

Exercises

1. Explain, using Ohm's Law, the reasons for the readings in Circuits 5 and 6.
2. If two resistors of 1000Ω and 1500Ω are in parallel, what is their combined resistance?
3. If two resistors of 1000Ω and 1500Ω are in series, what is their combined value?
4. If a circuit resistance of 500Ω allows 2 amps to flow, what voltage is being applied to the circuit?

Resistors in parallel (shunts)

1. Set up Circuit 1. Use an ammeter which will read up to 100 mA. Switch ON switch 1. Note the ammeter reading. Switch ON switch 2. The reading should increase approximately eleven times. This is because:

Switch 1 ON   —total resistance R = 1000Ω
Switches 1 and 2 ON—total resistance R is:

$$\frac{1}{R} = \frac{1}{1000} + \frac{1}{100} = \frac{11}{1000} = \frac{1}{90.9} \text{ so } R = 90.9\,\Omega$$

also:

Switch 1 ON—$I = \dfrac{V}{R} = \dfrac{4.5}{1000} = 4.5\,\text{mA}$

Switches 1 and 2 ON—$I = \dfrac{V}{R} = \dfrac{4.5}{90.9} = 49.5\,\text{mA}$

and $49.5 = 4.5 \times 11$.

2. Set up Circuit 2. Resistor 2 is to be such that the ammeter reading is increased exactly by 10 when both switches are ON as compared when switch 1 only is ON. The resistor R_2 can be made up from resistance wire. The calculations to find the value of R_2 are as follows. To obtain a 10-fold reading, the total circuit resistance must be lowered from 1000Ω to 100Ω when both switches are ON.

Thus $\dfrac{1}{100} = \dfrac{1}{1000} + \dfrac{1}{R_2}$

or $\dfrac{1}{R_2} = \dfrac{1}{100} - \dfrac{1}{1000}$

$= \dfrac{10-1}{1000} = \dfrac{9}{1000}$

$\therefore \dfrac{1}{R_2} = \dfrac{1}{111.1}$

$\therefore R_2 = 111.1\,\Omega$

Now, there is already 100Ω in series with R_2. R_2 will thus need to be 11.1Ω. Calculate the length of resistance wire to obtain 11.1Ω and add into the circuit.

Switch ON switch 1. Note reading. Switch ON switch 2. Note reading. The length of resistance wire in R_2 can be amended until the second reading is exactly ten times the first.

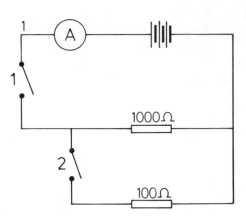

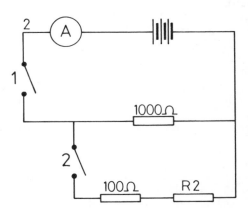

Note—Using carbon resistors which have a 10% tolerance error, an adjustment to the length of resistance wire needed to make R_2 will be necessary even if its length is calculated exactly. Calculations to find the length of resistance wire needed for a resistor of known resistance are simple because resistance wire is sold with different resistances to given lengths.

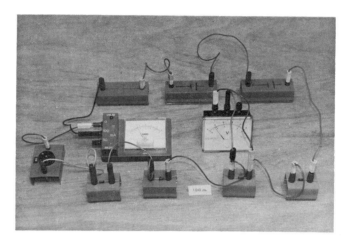

A photograph of Circuit 3

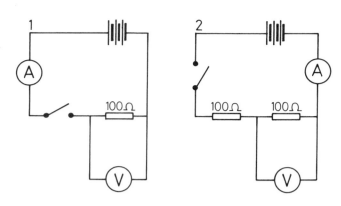

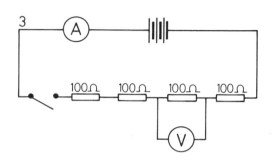

Voltage splitting

Resistors may be included in a circuit to change the current flowing in the circuit. The lower the resistance, the higher the current and the higher the resistance, the lower the current flowing.

Resistors may also be included in a circuit to alter the voltage in parts of the circuit.

1. Set up Circuit 1. 100 mA ammeter. 5 volt voltmeter. Switch ON. Note readings on ammeter and on voltmeter. Check that $V = IR$ (Ohm's Law).

2. Switch OFF. Add a second resistor as shown. Switch ON. Note readings on ammeter and voltmeter. Why is it that the voltage is now one half of what it was in Circuit 1?

Answer—There are a 4.5 V battery and two 100Ω resistors in series. By Ohm's Law $I = \dfrac{V}{R} = \dfrac{4.5}{200}$

= 22.5 mA. Thus 22.5 mA current is flowing through the circuit. The voltmeter is connected in *parallel* with ONE 100Ω resistor, through which 22.5 mA flows. Thus, by Ohm's Law

$$V = IR = \frac{22.5}{1000} \times 100 = 2.25 \, \text{V}.$$

3. Switch OFF. Add two more resistors as shown in the photograph and Circuit 3. Note readings on ammeter and voltmeter. Why is the reading on the voltmeter $\frac{1}{4}$ of what it was in Circuit 1? Explain by using Ohm's Law.

4. Set up Circuit 4. Switch ON. Why does the voltmeter give the full battery reading? The answer here lies in using your common sense and not in Ohm's Law.

Explain the reading on the ammeter by using Ohm's Law and your knowledge of resistors in parallel.

5. Switch OFF. Take out the voltmeter from the circuit and insert a switch. Switch ON switch 1. Note the reading on ammeter. Switch ON switch 2. Note reading on ammeter. Why is the second reading twice that of the first?

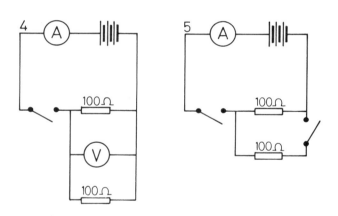

Exercises

1. Draw a circuit diagram which shows how a voltage of 25 volts may be drawn off from a circuit powered by a 50 volt battery.

2. Draw a circuit diagram showing how voltages in steps of 20 V at a time can be drawn off from a circuit connected to a 100 volt battery. You should show how to obtain 20 V; 40 V; 60 V and 80 V as well as 100 V.

3. If 12 amperes of current is flowing through a circuit of 10 ohms resistance, what voltage is applied to the circuit?

4. If a voltage of 250 V causes a flow of 2 amperes in a circuit, what is the circuit resistance?

Heating effect of electric current

Use a laboratory power pack to power some of the circuits shown on this page. Note the following when using a power pack.

1. Mains power switch on the mains socket must be OFF before plugging the power pack into the electric mains socket.

2. For the purposes of the circuits shown on this page, switches on the power pack should be positioned as follows before switching the mains switch ON:
a) Power pack switch OFF.
b) D.C.-A.C. switch at D.C.
c) 0–10; 10–20 switch at 0–10.
d) Voltage switch at 0.

1. Set up Circuit 1. Ammeter must be capable of reading up to 5 amperes. Switch power pack ON. Starting at 0 increase the voltage by 1 volt at a time up to 20 volts. Make notes of the ammeter and voltmeter readings at each voltage stage. Write down what happens to the Eureka wire at each stage. Switch OFF.

2. Set up Circuit 2. Ammeter must be capable of reading up to 5 amperes. Switch ON. Increase the voltage from 0 to 20 volts through 1 volt stages, taking readings at each stage. Write down exactly what happens, Switch OFF.

3. Set up Circuit 3. Ammeter must be capable of reading up to 5 amperes. Switch ON. Again take readings as the voltage is increased by 1 volt stages from 0 V to 20 V. Write down exactly what happens.

Note—Power is measured in watts.

watts = volts × amps.

4. Write down the power being consumed in experiments 2 and 3 at the moment just before the fuses melt.

To test for continuity in an electric circuit

1. Set up Circuit 4. The two wandering leads can be connected to a component to test whether its circuit is complete. This is the simplest possible of circuit testing devices. Use the circuit to test a variety of components.

2. Set up Circuit 5. This is a much more sensitive testing device than shown in Circuit 4. The larger the resistance in the circuit, the lower will be the meter reading. Why is this? Compare Circuit 4 and Circuit 5 across a 1000 ohms resistor. State what happens.

Exercises

Use equations based on Ohm's Law to solve the following:

1. If a full scale ammeter deflection shows in Circuit 5 when the two plugs are touched together, what resistance will be required in the circuit assuming the battery produces 5 volts and the ammeter resistance is 100 ohms and its maximum reading is 1 mA?

2. With a power supply of 250 V, what resistance is required in an electric fire which uses 1000 watts (1 kilowatt) of power?

3. With a power supply of 250 volts, what is the resistance of a 150 watt bulb?

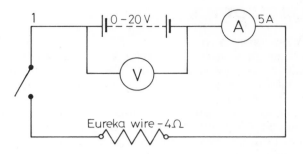

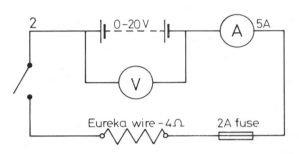

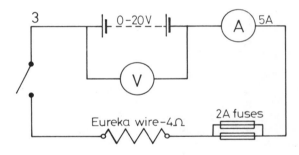

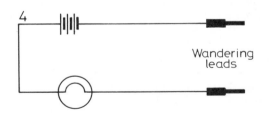

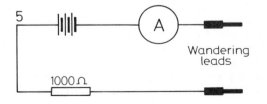

Electronics

Conductors and insulators

In a circuit consisting of a battery, copper wires and a bulb, the battery provides the electrical energy, the bulb transforms the electrical energy into light energy, and the copper wires allow electric current to flow between the battery and bulb. The copper wires have plastic outer coverings which do not allow electric current to flow through them. This outer covering protects the circuit from shorting.

The copper wire (and many other materials) which allows electric current to flow is a *conductor* of electricity. The plastic coverings (and many other materials) which do not allow electric current to flow are *insulators* against electricity.

Experiment

Construct Circuit 1. Place a variety of different materials between the crocodile clips. If the bulb lights when the switch is ON, the material is a conductor. If the bulb fails to light when the switch is ON, the material is an insulator.

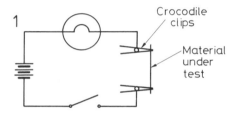

This experiment should show that many metals and some other materials such as carbon, are conductors. Rubber and many plastics are insulators. The structure of atoms of materials which conduct electric current are such that electrons can break away from the atom with ease. In fact an electric current can be thought of as electrons passing from atom to atom along a conductor.

The structure of atoms of insulator materials are such that electrons cannot easily break away from the atoms, thus an electric current cannot pass along, or through, an insulator.

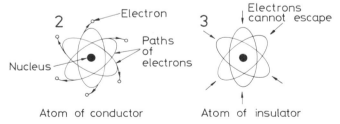

Theoretical diagrams of conductor and insulator atoms

Semiconductors

A third group of substances, of great importance in electronics, are known as *semiconductors*. Semiconductors are materials which cannot normally conduct electricity, but can be made to do so if encouraged.

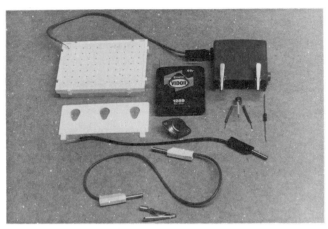

Some equipment for building test circuits. S-DeC and its front panel; components holder with crocodile clips; 4 mm plugs and leads; other components

The most commonly used semiconductor materials today are *silicon* and *germanium*. Silicon is possibly the best known because of the popular modern term 'silicon chip' and its association with computers and micro-electronics.

It might be thought that only pure silicon and germanium can be used for electronics. However it is only by the addition of minute quantities (as little as 1 part per million) of impurities to pure silicon or germanium that the high insulation properties of these materials can be controlled. Adding the impurities is also important in that they determine the type of semiconductors being produced. Examples of impurities used to *dope* silicon and germanium to produce semiconductor materials are—phosphorus, iridium and boron—among others.

Depending upon the material used for doping, two types of semiconductors are produced. These are known as n-type and p-type.

N-type semiconducting material

The doping in n-type material alters the structure of the material, creating an *excess* of electrons. N-type semiconductor material can be said to be negatively charged.

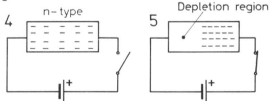

Theoretical circuit showing action of n-type semiconductor material

Circuits 4 and 5

These are theoretical circuits which cannot be constructed. When Circuit 4 is switched ON, the excess electrons in the n-type semiconductor material flow towards the positive side, creating an insulating barrier or *depletion region* on the negative side.

P-type semiconducting material

The doping of p-type material alters the structure of the material, creating a *shortage* of electrons. This leaves 'holes' in the material where the electrons should be. P-type semiconductor material can be said to be positively charged.

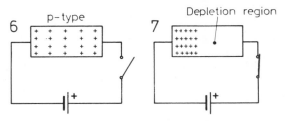

Theoretical circuit showing action of p-type semiconductor material

Circuits 6 and 7

When Circuit 6 is switched ON, electrons move from one 'hole' to another filling those on the negative side of the material, leaving a depletion region on the positive side—Drawing 7.

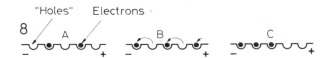

Theoretical diagram showing movements of electrons and 'holes'

The pn junction

On their own p-type or n-type semiconductors would act as insulators in a circuit and electric current would not flow through them. Joining p and n type semiconductors together however creates a pn junction and, depending on how this is connected in a circuit, current can flow through the junction.

Reverse bias

When the n-type material of a pn junction is connected to the positive side of a circuit no current will flow in the circuit because a double depletion region has been created.

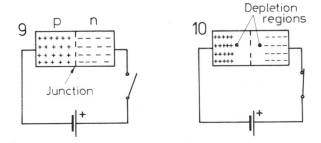

Reverse bias of a pn junction

Forward bias

When the n-type material of a pn junction is connected to the negative side of a circuit current can flow. The n and p materials compensate each other and no depletion region is formed.

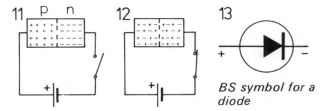

Forward bias of pn junction

Notes

1. Nearly all diodes are silicon junctions consisting of p and n type materials. Diodes are *uni-directional* and allow current to flow only in one direction.

2. Transistors are also p and n type semiconductor junctions made up in three layers in the form npn or pnp. Nearly all silicon transistors are npn and nearly all germanium transistors are pnp. Transistors have three connections—collector (c), emitter (e) and base (b). The semiconductor material connected to the collector connection is more heavily 'doped' than is the semiconductor material connected to the emitter. There is not sufficient space in this book to explain how the excess electron and 'hole' theory of semiconductor action can be applied to the functioning of a transistor. The reader is advised to consult other books. See bibliography, pages 127 and 128.

3. In Drawings 14 and 15 negative and positive connections are shown. It is important that the connectors are biased as shown in circuits which incorporate transistors. It should be noted that in an npn transistor the collector should be connected so as to be positive to the base, a larger positive voltage always being applied to the collector compared with the base. The emitter should, however, always be negative to both base and collector. Pnp transistors are biased differently as shown.

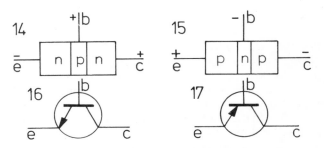

npn and pnp transistors and BS symbols

Diodes

The most common diodes in general use consist of junctions of p-type and n-type silicon semiconductor materials. Some germanium diodes are available, mainly for use in audio circuits. In general, the voltage drop across a silicon diode is several times higher than the voltage drop across a germanium diode.

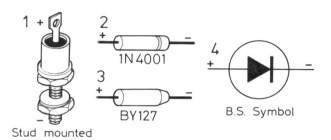

Stud mounted

Some diodes for use in school technology circuits

The diode type numbers are those given to components sold by Radio Spares Components Limited. The stud mounted diode is suitable for high power rectification (either 13 amperes or 26 amperes). 1N4001 is a plastic encapsulated diode capable of passing a maximum of 50 V. BY127 is a high voltage diode capable of passing up to 1250 V. A coloured band or shaped end identifies the n-type material. This is usually connected to the negative side (forward bias). The positive side of a diode is called the *anode* and the negative side the *cathode*.

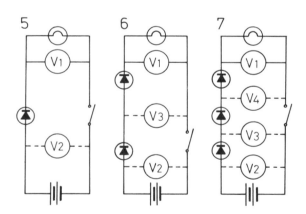

Voltage drop across a diode

Set up the circuits shown in Drawings 5, 6 and 7 in turn. Use 1N4001 diodes, a 3 V battery and a 2.5 V filament bulb. It is not necessary to use more than two voltmeters, one of which can be used to test at positions 2, 3 and 4 with the aid of 'flying leads'. Note the following:
a) Each voltmeter reading.
b) When bulb lights or fails to light.
c) Calculate the voltage across each diode—V2 – V1;

V2 – V3; V3 – V4. Is the voltage across each diode the same?
Note—Diodes are sometimes used to regulate circuit voltages.

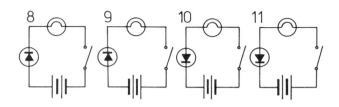

Forward or reverse bias

Set up the circuits shown in Drawings 8, 9, 10 and 11 in turn. Use a 1N4001 diode in each case.
(a) Note what happens in each case when the switch is ON.
(b) Which of the circuits are forward biased and which reverse biased?

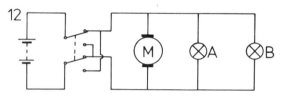

Modification of a reversing circuit

Set up the circuit shown by Drawing 12. Use batteries and bulbs to suit the voltage needed to drive the motor. Add diodes to the circuit so that when the motor revolves clockwise, bulb A lights and when the motor revolves anti-clockwise, bulb B lights. Test your circuits.

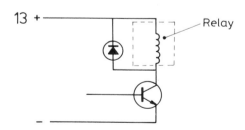

Diode in relay circuit

One use of diodes is to protect a circuit from a back current surge or back EMF (electromotive force) when relays are operated by transistors. In the circuit given (Drawing 13) is the diode forward, or reverse biased?

Light emitting diodes

Many circuits are designed in which a visual indication of circuit functioning is desirable. Filament bulbs could be used for this purpose, but light emitting diodes or LEDs are more suitable.

(a) LEDs are cheap, being only about twice the price of filament bulbs.
(b) LEDs emit far less heat than do filament bulbs.
(c) LEDs use far less current than do filament bulbs. In typical applications at similar voltages, LEDs use between 10 mA and 40 mA as against possibly 250 mA used by a filament bulb.
(d) LEDs are almost indestructible and thus have a far longer life than do filament bulbs. They are therefore more reliable in use.

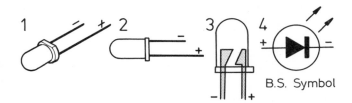

Light emitting diodes (LEDs)

LEDs are pn junctions, made from gallium phosphide and gallium arsenide phosphide. The junction is encased in coloured plastics, the hemispherical end of which forms a lens for the LED. To identify the anode (positive) look for the longer connector or for a flat on the rim of the body of the component.

Resistors in series with LEDs

LEDs normally work at between 2 and 2.5 volts. In order to obtain a correct working voltage for a LED in a circuit, a resistor will usually need to be placed in series with a LED. Ohm's Law can be used to calculate the resistance of the resistor. By Ohm's Law $V = IR$ where V = voltage, I = current, R = Resistance. Thus $R = \dfrac{V}{I}$.

Let V = circuit voltage, V_F the voltage required to operate the LED, and I_F the current required to operate the LED.

Then: $R = \dfrac{V - V_F}{I_F}$

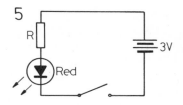

In Circuit 5, if $V_F = 2\,V$ and $I_F = 15\,mA$

then $R = \dfrac{3-2}{0.015} = \dfrac{1000}{15} = 66.6\,\Omega$

Exercise

1. Construct Circuit 5 using a resistor of the nearest preferred resistance (68 Ω). Does the LED light?
2. Replace the LED with one working at 2.5 V. Give reasons for what happens.
3. Reverse the positive and negative connections to the LED. What happens?

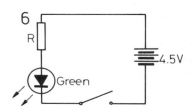

Construct the circuit shown in Drawing 6 using a green LED. This will require 2.0 V at 10 mA.
1. Calculate the size resistor needed in the circuit.
2. Construct and test the circuit.
3. Construct and test a similar circuit for a yellow LED working at 2.2 V and 15 mA in a circuit powered by a 6 V battery.

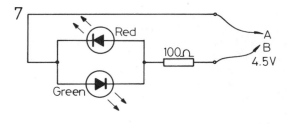

1. Construct the circuit shown by Drawing 7.
2. Use the circuit to check the polarity of a 4.5 V battery.
3. Which lead, A or B, is on the positive terminal when the red LED lights?

85

Transistors

The most important group of semiconductor devices are transistors. Transistors were first discovered/invented in 1948 by William Shockley, Walter Brittain and John Barden. All three received Nobel prizes for their work on transistors.

Since their inception transistors have revolutionised electronic and electrical applications.
1. Components and units have been drastically reduced in size.
2. Costs of components and units have considerably lessened.
3. Transistors are much more reliable than earlier electronic devices.

Transistors are available in a large variety of shapes and sizes, but all consist of npn or pnp junctions with three connectors—collector, emitter, base. Nearly all npn transistors are silicon and nearly all pnp transistors are germanium.

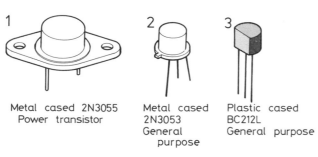

1. Metal cased 2N3055 Power transistor
2. Metal cased 2N3053 General purpose
3. Plastic cased BC212L General purpose

Full size drawings of transistors used in experiments shown in this book

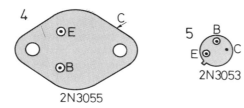

4. 2N3055 — ⊙E, C, ⊙B

5. 2N3053 — B, E, C

Full size drawings from underneath showing positions of connectors

Handling transistors

Usually all three connectors can be easily seen, but in the case of transistors such as 2N3055 (Drawing 1), the case is the collector connection, so only two connecting wires are visible. When handling small transistors the wire connectors must be treated with care. Note the following:
1. Collector, emitter and base connections must be identified before connecting a transistor into a circuit.
2. In any circuit the transistor leads must not touch one another.
3. When connecting an npn transistor, collector must be positive, emitter must be negative, and base must be biased positive (Drawing 6).

A 2N3055 power transistor

4. When connecting a pnp transistor, collector must be negative, emitter must be positive, and base must be biased negative (Drawing 7).

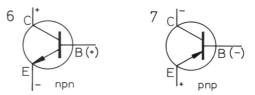

6. C +, B (+), E −, npn

7. C −, B (−), E +, pnp

Positive, negative and bias for e, b and c

5. Connections should never be bent close to the transistor case (Drawing 8).
6. When soldering connections to a transistor take care that the transistor itself is not heated. See details of 'heat sink' soldering, page 114.
7. If using S-DeCs for experimental circuits the wires of small transistors will require extending (Drawing 9).
8. When experimenting, coloured plastic sleeving slipped over the connecting wires will assist in identification of collector, emitter and base.

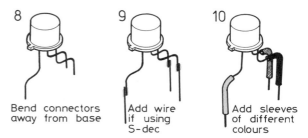

8. Bend connectors away from base

9. Add wire if using S-dec

10. Add sleeves of different colours

Precautions when handling transistors

9. Some of the experiments described here employ an ohmmeter. The ohms range should be set to $R \times 1$ if using a multi-meter. An ohmmeter incorporates a battery and care must be taken to note the positive and negative connections of this when connecting the instrument to a circuit.

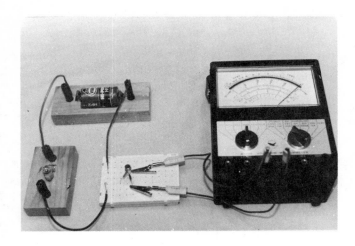

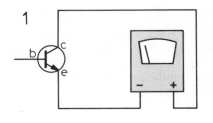

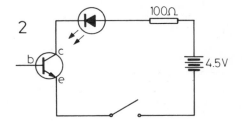

Experiments

For the experiments shown on this page use either a 2N3053 or a 2N3055 transistor (Radio Spares Components numbering). The ohmmeter should be set to $R \times 1$ (or its equivalent).

1. Construct the circuit shown by Drawing 1. Does the emitter to collector connections have a low or a high resistance?

2. Construct the circuit shown by Drawing 2. Does the LED. light when the switch is ON?

3. Construct the circuit shown by Drawing 3 and the photograph. Does the emitter to collector connection show a low or a high resistance? Does the filament bulb light? Now reverse the polarity of the 1.5 V battery. What happens? Can you explain the results of this experiment?

4. Construct the circuit shown by Drawing 4. Does the LED light? Does the filament bulb light? Now reverse the polarity of the 1.5 V battery. What happens? Can you explain the results of this experiment? Check the voltages by inserting voltmeters across the circuit as shown by broken lines. What readings are obtained? Can you explain these readings?

5. Construct the circuit shown by Drawing 5, but using either npn transistor 2N3053 or 2N3055 in place of the pnp transistor shown in Drawing 5. Switch ON and note what happens. Can you explain the results of this experiment? If you can obtain a suitable pnp transistor test this circuit with a pnp transistor. What happens when the switch is ON?

6. In Circuit 4, replace the 2.5 V filament bulb with a diode (1N4001). What happens? If the diode polarity has been correctly chosen it is being used as a 'voltage dropper'.

7. Using different npn transistors repeat each of the experimental circuits making notes of your results.

8. The word 'transistor' is derived from 'transferable resistor'. Explain how you think 'transferable resistor' describes a transistor.

9. Study Drawing 6. This drawing is an attempt at showing the electron flow in an npn transistor. Describe Drawing 6 in words.

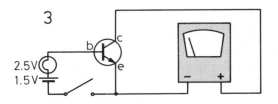

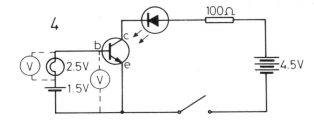

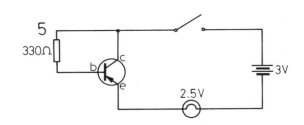

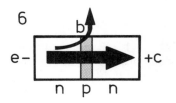

The transistor as a switch

Pages 82 to 87 have given some information about the operation of npn transistors. A small positive current applied to the base of an npn transistor causes a large current to flow through the emitter to collector connections. If the positive supply to the base is removed, the large flow of current through the emitter/collector ceases. The reasons for this are that if a positive bias is applied to the base there is a low resistance across the emitter/collector connections; with no bias at the base, the resistance across emitter/collector is high.

The circuits shown so far include a separate battery to provide the positive base bias. The positive side of the battery in the circuit connected to the emitter and collector can however be used to supply the positive bias to the base, by placing a resistor between the base and the positive pole of the circuit battery. The resistor effectively lowers the voltage at the base.

Experiments

1. Construct Circuit 1. Note that the resistor between the positive of the battery and the collector is 100 Ω, but the resistor between the positive supply and the base is 1000 Ω. When the circuit is switched ON, LED A lights, but LED B does not. This indicates that there is insufficient current passing through LED B for it to operate and that the transistor must be operating for LED A to light.

Remove the base connection. What happens? Change the 1000 Ω resistor to a 100 kΩ variable resistor. Adjust this variable resistor from zero through to 100 kΩ. Make notes of what happens.

2. Construct Circuit 2. Connect the flying leads first to A, and then to B, then to C. Make notes of what happens. The use of two resistors, one each side at A, B and C is known as *potential dividing*.

Explanation of results

The voltages available at A, B and C are found by applying the formula: Base voltage $= \dfrac{R_1}{R_1 + R_2} \times V$.

Thus:

(a) Flying leads not touching A, B or C, the base of the transistor is at zero volts. Emitter to collector has a high resistance. No current flows. LED does not light.

(b) At A. Base voltage $= \dfrac{10}{1+10} \times 4.5 = 4.1$ V

(c) At B. Base voltage $= \dfrac{4.7}{4.7+4.7} \times 4.5 = 2.25$ V

(d) At C. Base voltage $= \dfrac{1}{1+10} \times 4.5 = 0.41$ V

At A and B base +ve bias is sufficient to cause the transistor to work, the transistor switches ON, the LED lights. At C, 0.41 V is an insufficient base bias, the transistor does not work, the transistor switches OFF, the LED does not light.

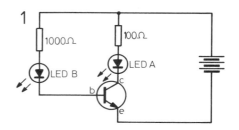

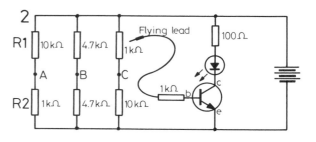

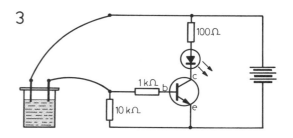

3. Construct Circuit 3. Also see photograph. This circuit shows an application of a transistor acting as a switch. When constructing this circuit note that the two flying leads go through a strip of plastic resting on the top of a glass container. The leads protrude through the plastic by about 25 mm.

Slowly fill the glass with water.

(a) Describe, in writing, what happens.
(b) Explain why the circuit functions as it does.
(c) What has happened to the base bias?
(d) Re-draw the circuit for a pnp transistor.

Note—The 10 kΩ resistor acts as a block, preventing the emitter of the transistor from becoming positive.

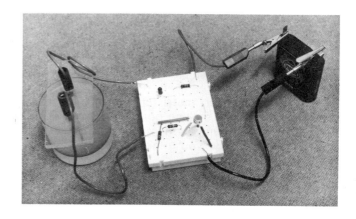

Light dependent resistors (LDR)

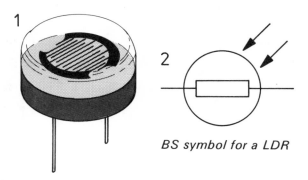

An ORP 12 light dependent resistor

Light dependent resistors—LDRs—can be fitted into circuits in a similar manner as are normal resistors. As the light falling on an LDR increases, so its resistance decreases. The LDR shown in circuits here is the ORP12 (Radio Spares Components Limited rating). In the dark the ORP12's resistance is 10 MΩ. This reduces to as little as 130 Ω as the light falling on it increases. Its resistance can vary from a maximum to a minimum in as little as 0.3 seconds. The ORP12 is a cadmium sulphide resistor set in a clear plastic.

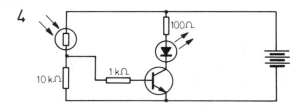

Experiments

1. Set up the circuit shown by Drawing 3. Set the ohmmeter to the $R \times 100$ range. Allow light to shine on to the LDR.
(a) Make a note of the ohmmeter reading.
(b) Shade the light by moving a hand between the LDR and the source of light. What happens to the reading on the meter?
(c) Cover the LDR so that it is completely in the dark. What happens?
(d) Reverse the polarity of the LDR in the circuit. Repeat (a), (b) and (c). What conclusion do you draw from the results?

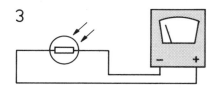

2. Construct the circuit shown by Drawing 4. The LDR is taking the place of a resistor in a potential divider circuit (see page 88).

(a) Shine a light on the LDR. What happens?
(b) Cover the LDR so that it is in the dark. What happens?
(c) Explain the results of (a) and (b) in writing.

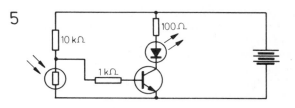

3. Construct the circuit shown by Drawing 5. This is the same circuit as shown by Drawing 4, with the LDR and 10 kΩ resistor reversed.
(a) Shine a light on the LDR. What happens?
(b) Cover the LDR so that it is in the dark. What happens?
(c) Explain the results of (a) and (b) in writing.

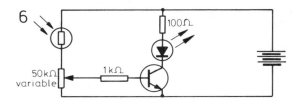

4. Construct the circuit shown by Drawing 6. Adjust the 50 kΩ variable resistor so that:
(a) The circuit is very sensitive to changes of light falling on the LDR.
(b) The circuit is least sensitive to changes of light falling on the LDR.
(c) Reverse the LDR and the 50 kΩ variable resistor. Experiment with various settings of the variable resistor and changes of light falling on the LDR. Explain the results of your experimentation in writing.

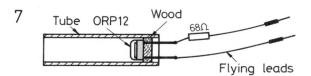

LDR set in a tube

LDR housing
A housing in which an ORP12 LDR can be fitted can be made from ∅ 18 mm plastic tubing. See Drawing 7. With the LDR housed in this manner, it can be pointed at a light source without having to take reflected light into account. A torch can also be shone into the tube from a distance thus signalling a transistor to switch.

Thermistors

Two types of thermistor are made. These are positive temperature coefficient (ptc) types and negative temperature coefficient (ntc). The experiments shown here use a TH3 ntc rod thermistor (Radio Spares Components Limited type numbering). The resistance of the TH3 thermistor is approximately $380\,\Omega$ at 25°C lowering to $28\,\Omega$ when hot. It is a general purpose current limiting thermistor designed mainly for circuit protection.

As the temperature applied to an ntc thermistor increases so its resistance to the flow of electrical current decreases. Thermistors are therefore temperature dependent resistors or TDRs.

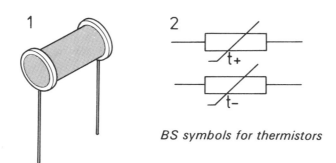

Rod thermistor TH3

BS symbols for thermistors

Thermistors are made by firing a compressed mixture of metal oxides such as those of manganese, nickel and cobalt.

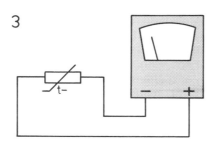

Experiments

1. (a) Construct the circuit shown in Drawing 3. Set the ohmmeter to its $R \times 100$ range. Heat the thermistor (a lighted match or candle is suitable). Note the meter reading.
(b) Cool the TDR by placing it into a container with ice. Note the meter reading.
(c) Reverse the polarity of the connections between the TDR and the ohmmeter. Repeat (a) and (b). Note the meter readings.
(d) What readings do you think you would obtain if a positive temperature coefficient (ptc) thermistor was placed in the circuit (Drawing 3) in place of the ntc thermistor?

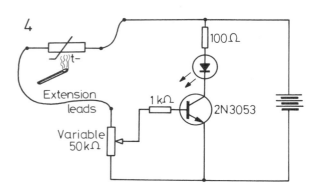

2. Construct the circuit shown by Drawing 4 and the photograph. Switch ON. Adjust the $50\,k\Omega$ variable resistor so that the LED is not alight. Hold a lighted match under the TDR for a few seconds. Note what happens. Explain in writing what happens within the circuit as a result of this experiment.

Exercises

1. Design a circuit which can be used in an ice warning device. Construct the circuit you have designed.
2. Devise a method of calibrating the device for which you have designed an ice warning circuit.
3. Find out about:
(a) Bead thermistors
(b) Disc thermistors
(c) Thermistor probes.

Note

The two circuits shown in Drawings 3 and 4 are not very sensitive. Better results would be obtained if the variation of current flow in the circuit resulting from changes in temperature of the transistor, could be amplified.

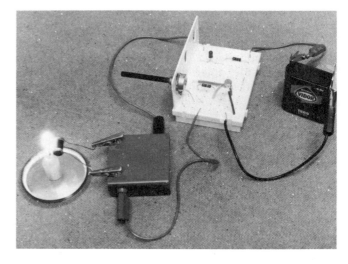

A photograph showing circuit 4

Transistor amplification

The word 'amplifier' is in common use. 'Amplifiers' are used, for example, with musical instruments such as an electrical guitar. The small electrical 'signals' produced by the electronics circuit in the instrument are 'amplified' or made into very much stronger 'signals' by an 'amplifier'. A loudspeaker connected to the amplifier enables the stronger 'signals' to be clearly heard. This is shown in Drawing 1.

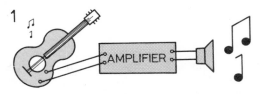

The output signal via the loudspeaker is much greater than the input signal from the instrument's electronics circuit.

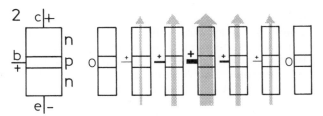

Diagrammatic description of transistor current gain as base bias current is increased.

Current amplification in a transistor can be shown diagrammatically as in Drawing 2. As a small current flowing at the base of the transistor increases, so a larger current flows through the emitter to the collector. As the small base current decreases, so does the larger emitter/collector flow. Variations of base current therefore control an amplified emitter/collector current.

Experiments

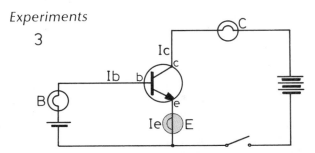

Earlier experiments with transistor circuits (pages 86 to 88) show that a transistor can be used to amplify current.

1. Construct the circuit shown by Drawing 3. This circuit is similar to Drawing 4 on page 87. The LED has, however, been replaced with a filament bulb. Switch ON. Bulb B will not light. Bulb C will light. The positive bias, with its small current flow at the base of the transistor is causing a larger current to flow through the emitter/collector connections.

Note

Kirchoff's First Law can be demonstrated here. This states that the current flow at c (I_c) plus the current flow at b (I_b) equals the current flow at e (I_e). $I_c + I_b = I_e$. Connect a bulb at the position shown shaded (E) in Drawing 3. This bulb should show brighter than that at C.

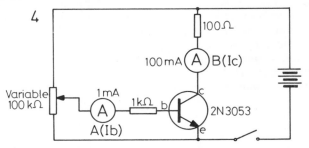

2. Construct the circuit shown by Drawing 4 and the photograph. Note that ammeter A should read up to 1 mA and therefore gives readings in μA (microamperes) and ammeter B should read up to 100 mA and therefore gives readings in mA (milliamperes). Switch ON. Adjust the 100 kΩ variable resistor until a maximum reading shows on ammeter B.
Typical readings then might be:
Base current (I_b) = 450 μA (0.45 mA)
Collector current (I_c) = 42 mA.

The ratio $\frac{I_c}{I_b}$ = the current gain of the transistor. The symbol for transistor current gain is h_{FE}. In this example—

$$h_{FE} = \frac{I_c}{I_b} = \frac{42}{0.45} = 93.3$$

Note that this current gain is for this transistor in this circuit. The 2N3053 can produce gains of from 50 to 250 depending upon the circuit conditions under which it is operating.
3. Construct the circuit shown in Drawing 3 using other transistors, e.g. 2N3055. Calculate the gains of these transistors for the given circuit conditions.

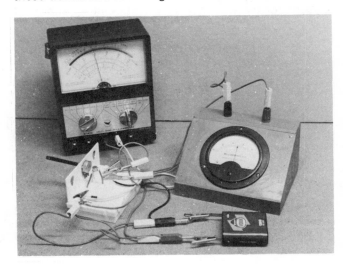

The Darlington Pair amplifier

As has been shown in earlier pages, transistors can be incorporated in circuits to amplify current flow. A small positive current applied to the base of an npn working transistor is amplified into a much larger current flow through the emitter/collector connections. If the amplified current from one transistor is fed into the base of a second transistor, the amplification of current is considerably greater, being the product of the current gain of the two transistors. Taking the current gain of a 2N3053 transistor as being 100 (in a particular circuit), if two transistors are coupled as shown (in Drawing 1), the total gain of the two devices is $100 \times 100 = 10\,000$.

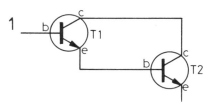

Drawing 1 shows the principle of this coupling of two transistors—known as a *Darlington Pair*. Note that the emitter of T1 is taken to the base of T2 and that the collector of T1 is taken to the collector of T2.

Experiments

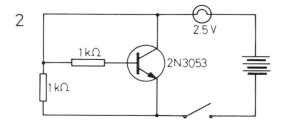

1. Construct the circuit shown by Drawing 2. Switch ON. Note what happens. Write an explanation giving reasons for the way in which the circuit functions.

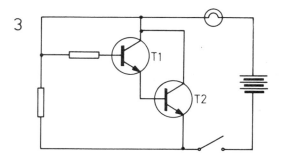

2. Construct the circuit shown by Drawing 3 by adding a second transistor as shown. Switch ON. Note what happens. Write an explanation giving reasons for the way in which this circuit functions.

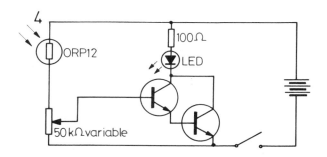

3. Construct the circuit shown by Drawing 4. Switch ON. Adjust the 50 kΩ variable resistor as necessary. Note how sensitive this circuit is to light falling on the light dependent resistor ORP12.

Write an explanation describing the functioning of this circuit.

Change the positions of the ORP12 and the 50 kΩ variable resistor. Adjust the variable resistor. Note what happens. Write an explanation giving reasons for the way in which this circuit functions.

Replace the ORP12 with a thermistor TH3. Adjust the 50 kΩ variable resistor as necessary. Apply a lighted match to the TH3. Note and explain what happens.

Exercise on experiment 3
Design and construct a circuit which will detect small quantities of water in a dish of common salt.

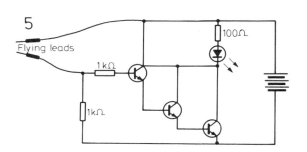

4. Construct the circuit shown by Drawing 5. Put one flying lead in one ear and the second flying lead in the other. Note what happens.

Hold hands with a number of people in line, the first and last in the line each holding a flying lead. How many people can hold hands in line and the circuit still work?

This circuit can also be used to detect electrical current flow in leaves, pieces of coal, wood shavings etc.

Draw a line on paper with an HB pencil. Put one flying lead at one end of this line and move the second flying lead along the line towards the first. Note what happens. *Note*—The current gain of this circuit could be as much as $100 \times 100 \times 100 = 1\,000\,000$.

Integrated circuits

The type of circuit construction used so far is known as *discrete*. This word describes circuits made up from individual components. The components used have been *active* e.g. diodes and transistors or *passive* resistors.

Integrated circuits are complete circuits within themselves. The discovery of the transistor soon led to the discovery that two transistors could be built in the same piece of semiconductor material. Then it became possible to include diodes, resistors and capacitors all within the same piece of semiconductor material. Modern methods of production have brought about a completely new technology—microelectronics, using 'silicon chips'—integrated circuits contained in very small 'chips' of silicon. Such 'chips' may be as small as 2 mm square yet contain numerous components. Their cost is low and they are extremely reliable when used in circuits. Each 'silicon chip' is mounted in a plastic case and its connections are taken to pins set in the sides of the case.

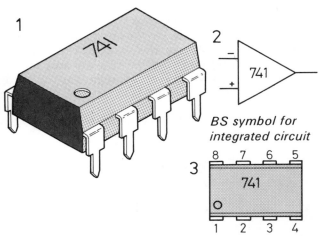

BS symbol for integrated circuit

Numbering of connections

The 741 operational amplifier integrated circuit (4x scale)

One integrated circuit of value in school technology is the 741 operational amplifier (op. amp.). This is an 8-pin, dual-in-line integrated circuit. With its casing this 'chip' measures 9 mm by 7 mm by 3 mm. It contains over 30 components, including 20 transistors. Its gain can be as high as 10 000. Its pins are numbered as shown in Drawing 3—note the small hole at 1—and then anti-clockwise to number 8. Correct pin connections are very important. Initially, note that pin 4 is connected to negative supply and pin 7 to positive supply.

Experiment 1
Construct the circuit shown by Drawing 4. Adjust the 2kΩ variable resistor until the LED is OFF when the ORP12 faces a light source.

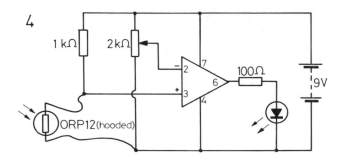

(a) Shade the ORP12. What happens?
(b) Point the ORP12 at a light source; move a hand over the ORP12. How sensitive is this circuit compared with that shown on page 89 which operated with a single transistor?
(c) Reverse the diode. Repeat (b). What happens?
(d) Explain how the circuit could be amended so that the LED is ON when the ORP12 is pointed at a light source.

Zero volts
'Zero volts' (0V) can be achieved by connecting two batteries in series, the zero connection being taken from the central positive and negative junction.

This junction is neutral and can thus be either postive or negative with respect to the two end terminals to give two different circuits: positive to 0 volts; negative to 0 volts. The pins 2 and 3 of a 741 op. amp. are voltage input pins—they are the *inverting input* (pin 2) and the *non-inverting input* (pin 3). The differential between these 2 pins (V2-V1) is amplified as *output voltage* at pin 6. The output voltage can therefore be either positive or negative.

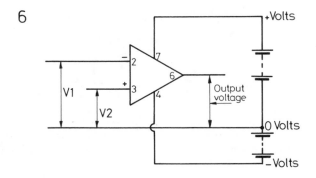

With this in mind it should be possible to construct the circuit shown in Drawing 4 so that it will cause two diodes (LEDs) placed opposite to each other in the circuit, to function.

7

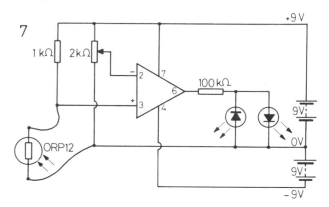

8

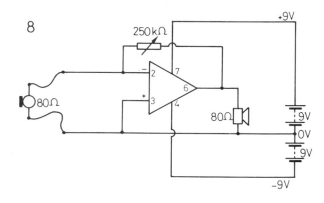

Experiment 2

Construct the circuit shown by Drawing 7, and the photograph. Use different coloured LEDs. Adjust the 2 kΩ variable resistor as necessary.

(a) Test the circuit as for Experiment 1.
(b) Describe, in writing, what happens when the circuit is tested. This circuit is known as a 'differential amplifier'.
(c) Describe, with the aid of circuit diagrams, how a circuit for a burglar alarm could be constructed based on the circuit shown by Drawing 7.
(d) Find out and describe how to fit a relay in 741 op. amp. circuit.

Amplifiers do amplify—Experiment 3

Construct the circuit shown by Drawing 8. Note the variable resistor between input terminal 2 and output terminal 6. Some of the output is thus modifying the input. This is known as *negative feedback*. Note that pin 2 is shown as negative and pin 3 as positive. In this circuit pin 3, the *non-inverting input*, is connected to 0 volts. Pin 2 is connected to the input and the voltage at pin 2 is a variable negative voltage. This circuit is known as an *inverting amplifier*. The variable 250 kΩ resistor allows a variety of microphones or loudspeakers of different resistances to be used, other than the 80 Ω types shown.

Test the circuit. Adjust the variable resistor to obtain the best results.

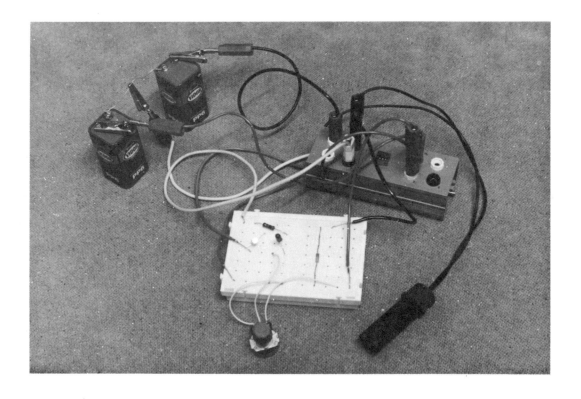

Computers

The IBM 5100 Portable computer. This is a compact portable computer designed for personal use. It fits easily on an office desk and can be used in an office, conference room, laboratory, job site or at home. Photograph by courtesy of IBM United Kingdom Limited.

An ICL-PERQ personal, scientific and engineering workstation, installed at the Science and Engineering Research Council's Rutherford Laboratory at Chilton, Oxfordshire. The picture illustrates the graphics facilities of the PERQ system and small, compact dimensions. Photograph by courtesy of International Computers Limited.

13 mm steel plate being cut to shape by oxy-acetylene cutting heads under water. Profile of the outline of the shape controlled by computer from tape-stored information. Photograph taken at J. C. Bamford (Excavators) Limited of Rocester.

Mechanical computers have been in existence for a considerable time. A mechanical calculator invented in 1642 could be said to be a computer. Various mechanical devices used for automatically controlling machinery in the 19th century could also be classed as computers. Computers depending upon electrical circuits were designed during the 1940s. With the advent of transistors, computers could be designed with semiconductor electronics circuits. Small size computers— microcomputers—were made possible with the introduction of 'silicon chips', which contain complex electronics circuits on tiny 'chips' of silicon, some as small as 2 mm or 3 mm square, most no larger than 6 mm square. Modern microcomputers, such as are found in schools and colleges, can process as much information as could computers of considerably larger sizes which depended upon semi-conductor circuits, prior to the introduction of silicon chips.

Two groups of computer are currently in use-digital computers and analogue computers. Here we are only concerned with digital computers. Analogue computers are used mainly for the processing of continuous information, such as the recording and analysis of weather conditions — temperature, pressure, rainfall, sunlight and other processes of a continuous nature. No matter how small, or how large, digital computers are nothing other than devices which can process arithmetical operations in a binary code of electrical current pulses.

Inputs

Instructions can be put into (input) a computer by a variety of methods:

Keyboard

A keyboard allows humans to 'communicate' instructions to computers by typing. Instructions typed on a keyboard attached to a computer will be processed within the computer.

Tapes

Plastic tapes with coatings of magnetisable material are used to feed instructions into computers. The common audio cassette tapes can be used for this purpose as well as such tapes made specifically for computers. Another tape used for computer input is punched paper tape with holes across in binary codes. Punched paper tape is often used in printing.

Punched cards

Punched cards containing information in holes punched across the cards in binary code are another means of feeding instructions into a computer.

Discs

The common disc for microcomputers is the 'floppy' disc. Other, larger discs are used with other forms of computer. Discs (similar to music records) for graphics display programs are currently under development. Information on such discs—again plastics with magnetised coatings—can be fed into a computer.

Other input devices

Other input devices such as document readers (for some forms of printing, for 'reading' cheques etc.), graphics light pens for amending graphics displays on a VDU and links to other computers via terminals are among other input devices.

Output

A variety of methods are employed for obtaining the output resulting from the processing of information fed into a computer.

Visual display unit

Commonly known as VDU. The output is displayed on a cathode ray tube screen, either as print or as graphics.

Printer

The output can be printed on paper by a printer of which several types are available. The output from a computer can be stored on say, a floppy disc, which can then be fed into a computer for printing via the computer, on a printer.

Typewriter

Some computers can be linked to a typewriter so that it automatically types the information obtained from the output.

Discs and tapes

The output can be stored on discs and/or tapes and output so stored can be retrieved as and when required.

Interfaces

In order to operate, e.g. a mechanical robot, the tiny output signals from a computer need to be amplified to provide a sufficiently large power source to drive a mechanical device. Interfaces must be designed for each particular mechanical device being operated. As will be explained later (page 98) the function of an interface is more complex than to act just as an amplifier.

The microprocessing unit (MPU)

The 'heart' of any computer is its processing unit. In microcomputers this is known as a microprocessing unit (MPU). MPUs consist of a number of silicon chips with associated circuitry, which deal with the information received via input devices and process it, as instructed, for passing to output devices. The information passes through the MPU circuits as a series of pulses of electrical current in a binary code.

The F100-L Microprocessor Chip, one of the world's most advanced 16 bit microprocessors. This is a 6 mm by 6 mm square chip containing 9000 electronic components, having a total length of 2.5 metres. The photograph is approximately 600 times the size of the actual chip.

Photograph by courtesy of Ferranti Electronics Limited.

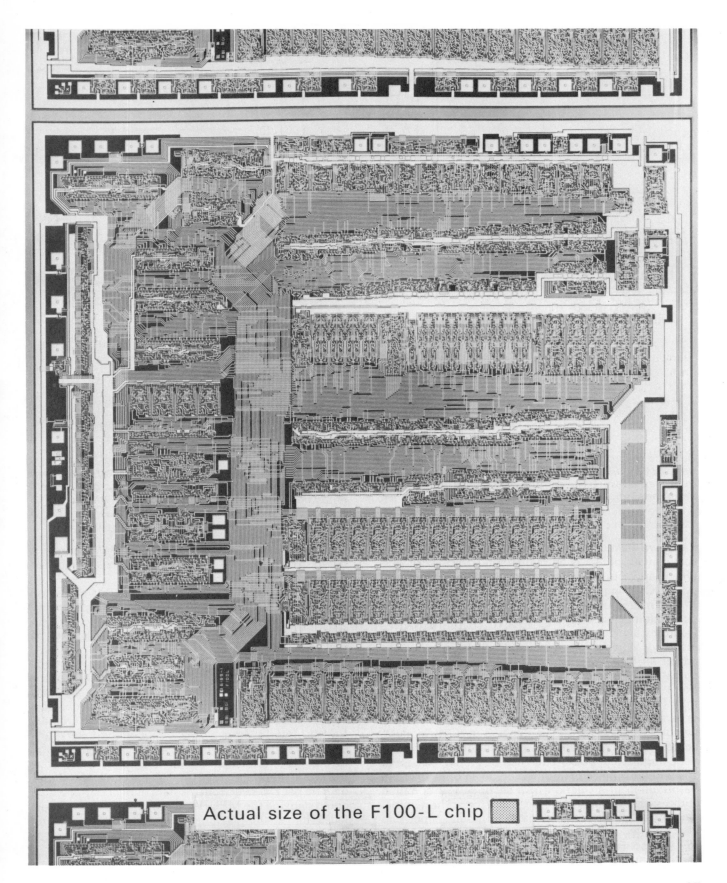

Actual size of the F100-L chip

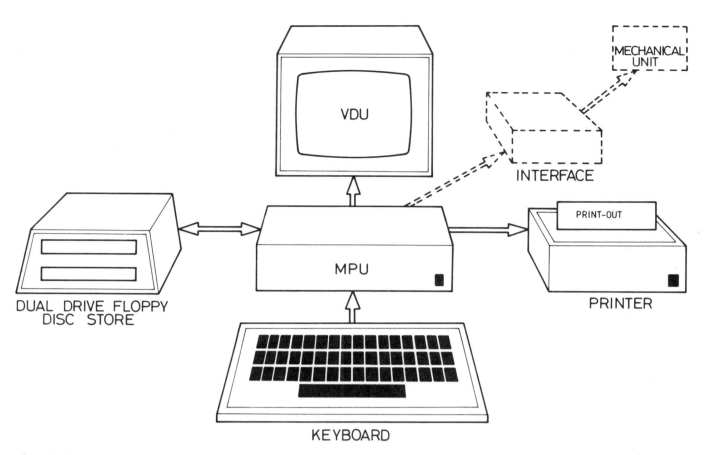

A typical microcomputer system. A microprocessing unit (MPU) with a keyboard. Peripherals shown are—a dual drive floppy disc storage unit; a visual display unit (VDU); a printer. An interface amplifier/buffer/decoder is shown in broken lines. Other peripherals may be added e.g. a computer aided design unit (CAD). Some microcomputers combine keyboard, MPU and VDU into what appears to be a single unit. Others combine the keyboard with an MPU.

Keyboard—similar to a typewriter keyboard with the same layout of letters and figures, but with extra keys which apply only to the operation of the microcomputer. One important extra key is the *return* key.

MPU—microprocessing unit. The MPU will contain some information storage capacity in silicon chips—RAM (random access memory) plus ROM (reading only memory). When electrical power to the MPU is switched off all RAM is erased, but ROM is retained. The data in ROM is said to be 'burnt-in' to the silicon chips.

Store—most microcomputers store information other than ROM and RAM on floppy discs, usually in dual drive units with mechanical drive. A floppy disc is a flexible plastic disc of diameter 135 mm (5¼ inches), coated with a magnetisable material on which magnetic impulses in binary code can be stored. Access to any particular piece of information on a floppy disc can be obtained in a fraction of a second. The common audio cassette tapes can be used for storing information but access to information on such tapes can take minutes rather than seconds.

VDU—visual display unit. Similar to a television screen (which can be used as a VDU). Whatever is typed on the keyboard is displayed on the VDU. Whatever is typed can be stored on the floppy discs and recalled (returned) for display on the VDU.

Printer—what is displayed on the VDU or stored on floppy discs (or tapes) can be printed on paper by the printer. Either print or graphics can be so printed. Continuous paper supply. The printer is only used as and when required.

Interface—the output of an MPU is in binary code signals usually through 8 sockets set in the rear of the unit. The power (current × voltage) of these signals is very small. In order to use the MPU as a mechanical control medium it is necessary to decode the binary signals and to amplify the decoded and very tiny electrical signals into a power supply sufficiently large to operate the mechanical item which is to be controlled. An interface system must also block any surges of 'feed back' power from the item under control which could damage the MPU. With properly designed interfaces, items such as vehicles, cutting machinery and other mechanical devices, can be controlled by programmes from the store of a microcomputer.

Above
A board of memory chips taken from the 380Z micro-processing unit.

Left
A 380Z microprocessing unit with keyboard. The dual drive floppy disc unit is part of this Research Machines computer, being housed on the left hand side of the casing.

Below
A Commodore PET microcomputer. In this example the microprocessing unit is contained in the same housing as the keyboard. The PET is connected to four peripherals—a VDU, a dual drive floppy disc unit, a printer and an interface. The interface is designed to allow a programme on a floppy disc to control and steer via the microprocessor a vehicle attached by a lead to the interface.

Binary, bits and pulses

Microcomputers function through electronics circuits contained in silicon chips within their microprocessors. In general it can be said that the circuits are made up as *two-state* processing devices—the devices can either *pass* a pulse of electric current or *not pass* a pulse of current. Such two-state devices are said to be *bistable*. A mathematical system which can be used to represent the bistable action of the processing devices is the *binary* system. The binary system is based on two digits only—1 and 0. When a processing device in an electronic circuit passes a pulse of current that pulse can be represented by the digit 1. No pulse passing can be represented by the digit 0. The digits 1 and 0 of the binary system are known as *bits* in computer terms. When a computer is said to be capable of storing or processing so many bits, it must be remembered that it is not 1s and 0s that are being stored or processed, but that the circuits involved will allow pulses of current to flow or not to flow up to the quoted storage capacity. A group of bits is a *byte* (or a 'word'). A byte usually consists of 8 bits but can, in some computers, consist of 16 or 24 bits. Computer storage capacity is quoted in so many 'K' of bytes, 1 K being 1024. A 4K store will be capable of dealing with $4 \times 1024 = 4096$ bytes (usually each byte being 8 bits).

Thus, in binary, a pulse of current flowing through a computer processing device can be represented by the digit 1. No pulse can be represented by the digit 0. See Drawing 1.

1

Pulse of current—1   No pulse—0

From 6 bits in line, 64 characters can be formed. One commonly used computer language is BASIC. The characters for working in BASIC are 64 in number.

Some examples of 6 bit groups to obtain 64 characters are shown in binary digits in Drawing 2.

2

0	0 0 0 0 0 0	A	1 0 0 0 0 1
1	0 0 0 0 0 1	B	1 0 0 0 1 0
2	0 0 0 0 1 0	C	1 0 0 0 1 1
3	0 0 0 0 1 1	D	1 0 0 1 0 0
4	0 0 0 1 0 0	E	1 0 0 1 0 1
5	0 0 0 1 0 1	F	1 0 0 1 1 0
6	0 0 0 1 1 0	G	1 0 0 1 1 1
7	0 0 0 1 1 1	H	1 0 1 0 0 0

As can be seen from the examples in Drawing 2, the 64 characters can each have a distinct binary code.

In the processing devices of a microprocessor unit the 64 characters can be recognised as pulses of current. See Drawing 3. Each group of 6 pulse/no pulse flows is clearly distinct from others in the set of 64.

3

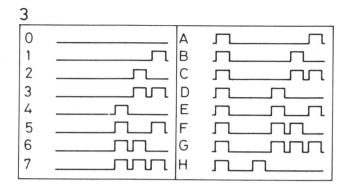

In this way, each of the 64 characters employed in a language such as BASIC, can be passed through the circuits of a microprocessor as a number of distinct, different and recognisable pulses of electrical current. Similarly a magnetisable tape or a floppy disc can be magnetised so as to produce 'messages' which the microcomputer can convert through its microprocessor and peripherals into characters displayed on a VDU or as a print-out in characters on a printer, or as a variable pulse current from which an interface can operate a mechanical device.

In the case of ROM (read only memory) of the microprocessor the circuitry of ROM is designed so as to permanently preserve 'messages' for delivery as electric pulses as and when required. These 'messages' can be delivered on orders from a keyboard, or from a punched tape or from other programmes.

Remember that the current pulses causing the microprocessor unit to function are passing through the circuits of the unit at a speed of 300 000 kilometres per second. 'Retrieval' of characters from ROM or say from a floppy disc, thus appear to be practically instantaneous.

Logic gates

In a microprocessor unit, the pulses of current (the bits)—which we have represented by binary digits—are controlled by *logic elements* within the electronic circuits of the unit. Logic elements are electronic circuits which act as switches. Because they are electronic switches, they have no moving parts. They are either ON to allow current to pass (represented by digit 1), or OFF, so not allowing current to pass (represented by digit 0). To rephrase the last sentence using different words—current can pass (OPEN) or not pass (CLOSED). Open/closed suggests gates, hence the term *logic gates*. Gate OPEN represented by digit 1. Gate CLOSED represented by digit 0.

Five logic gates are shown here. Three of these are common gates known as AND, OR and NOT. Combinations of these three common gates give a further two known as NAND and NOR. In order to show the link between logic gates and binary digits, *truth tables* can be written to demonstrate the functioning of logic elements as they open and close to allow pulses of current to pass or not to pass.

Circuits to demonstrate the action of logic gates

Logic gates in microprocessor units are within the circuits of the silicon chips making up the units. To demonstrate the action of logic gates TTL (transistor–transistor logic) silicon chips can be connected into circuits as shown in circuit diagrams on this page and on page 102 overleaf. A whole series of TTL chips can be purchased each with its own logic circuit. For the purposes of the circuits shown here a TTL-7400 (Radio Spares Components numbering) chip is used. This is a 14 pin quadruple 2-input NAND gate chip. With the aid of five different circuits, this chip can show the action of the five logic gates—AND, OR, NOT, NAND and NOR.

Symbols

Two groups of symbols are in use to draw logic gates. Both groups are shown next to the drawings of each of the five circuits.

Truth tables

Truth tables show how logic gates can be associated to binary digits. In the circuits shown here digit 1 shows the gate open—input switch ON, output light ON; digit 0 shows the gate closed—input switches OFF, output light (X) OFF. A truth table for each of the gates is shown against each of the five circuit drawings.

Drawing 1.

A plan of the TTL-7400 chip, showing the numbering of its 14 pins. The symbols showing the 4 NAND gate connections to the pins are not drawn on the actual chip casing. Pin 14 (Vcc) is the positive supply pin. Pin 7 (Gnd) is the 0 V pin, connected in the circuits shown to the negative power supply.

Drawing 2

Circuit for a two-input AND gate using the TTL-7400 chip. Both symbols for a two-input AND gate. Truth table. X is the output. From the truth table it can be seen that when A *and* B are ON, the LED (X) is ON. Thus A AND B = X.

Drawing 3

Circuit for a two-input OR gate using the TTL-7400 chip. Both symbols for a two-input OR gate. Truth table. From the truth table it can be seen that when A *or* B are ON, the LED (X) is ON. Thus A OR B = X.

Drawing 4

Circuit for a NOT gate using the TTL-7400 chip. Both symbols for a NOT gate. Truth table. From the truth table it can be seen that when A is ON the LED (X) is OFF; when A is OFF, the LED (X) is ON. Thus A NOT X.

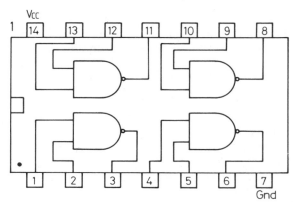

Plan of a TTL-7400 silicon chip—quadruple 2-input NAND gate chip

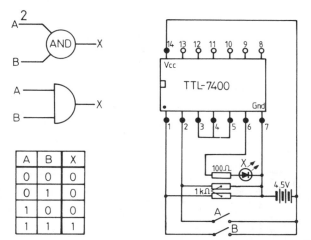

Symbols, truth table and circuit diagram for an AND gate

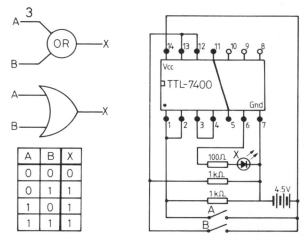

Symbols, truth table and circuit diagram for an OR gate

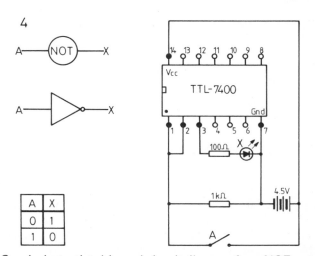

Symbols, truth table and circuit diagram for a NOT gate

Drawing 5
A NAND gate operates as if a NOT gate follows on from AND gate. The drawing shows a two-input NAND gate.

Drawing 6
Circuit for a two-input NAND gate using a TTL-7400 chip. Both symbols for a NAND gate. Truth table.

Drawing 7
A NOR gate operates as if a NOT gate follows on from an OR gate. The drawing shows a two-input NOR gate.

Drawing 8
Circuit for a two-input NOR gate using a TTL-7400 chip. Both symbols for a NOR gate. Truth table.

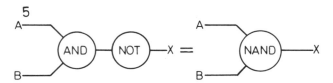

A NAND gate operates as if a NOT gate follows an AND gate

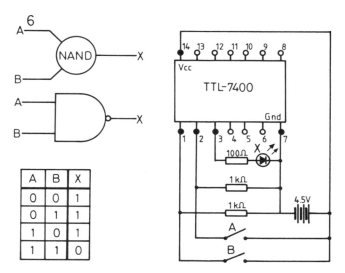

A	B	X
0	0	1
0	1	1
1	0	1
1	1	0

Symbols, truth table and circuit diagram for a NAND gate

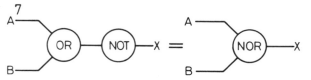

A NOR gate operates as if a NOT gate follows an OR gate

Note
In the ROM (read only memory) and RAM (random access memory) of a microprocessor unit (MPU), the silicon chips will contain complicated combinations of the logic gate circuits to process the pulses of current passing through the MPU.

Exercises
Construct each of the five gate circuits shown on this page and on page 101 overleaf. Check that the circuits operate as shown in the truth tables given with the circuit diagrams of the five gates.

An outline of the manufacture of silicon chips
Most silicon chip circuits are formed in layers in the chip. An outline of the processes involved follows procedures such as the following:
1. Pure silicon, 'doped' to form p-type semi-conductor material, is produced in cylinders of 100 mm diameter.
2. The cylinders are cut into wafers of about $\frac{1}{2}$ mm thick.
3. The chip circuits are designed, aided by computer aided design (CAD) equipment on a very large scale, several hundred times that of the finished chip.
4. The circuit designs are stored as computer programmes.
5. A surface coating of silicon dioxide is formed on the doped silicon wafers.
6. The circuit designs are printed, by photographic methods, several hundred to each wafer, at chip size. This printing is in photo resist material. Etching removes the silicon dioxide film not protected by the photo resist material.
7. Parts of each chip are re-doped, with possibly phosphorus, to form n-type semi-conductor material.
8. A second, and perhaps a third, layer of circuitry are built up on the chips in the same manner.
9. The chips are coated with aluminium.
10. By photo/resist/etching methods, parts of the aluminium are removed to form contacts to the circuits on the chips.
11. The chips are cut from the wafer into individual components.
12. The individual chips are mounted in plastic with the contacts connected to pins ready for use.

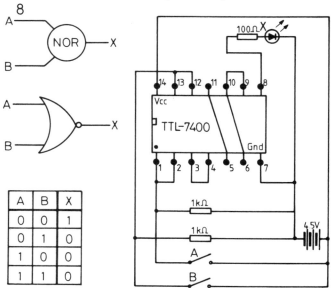

A	B	X
0	0	1
0	1	0
1	0	0
1	1	0

Symbols, truth table and circuit diagram for a NOR gate

Program for Morse Code translator

This program was designed to allow messages typed on the keyboard of an Apple 2 microcomputer to be translated into morse code produced as buzzes by the audio facility of the Apple 2. The program is in APPLESOFT BASIC.

The program was designed, tested and amended by Martin Cooper while a pupil at the Thomas Alleyne's High School, Uttoxeter.

```
]LIST

10   REM *** INITIALIZATION ***
20   READ N
25   DIM X$(N,2)
30   FOR I = 1 TO N
35   READ X$(I,1),X$(I,2)
40   NEXT
50   INPUT "TYPE YOUR MESSAGE ";A$
60   PRINT
70   REM  *** START OF MESSAGE CHARACTER ***
80   REM
90   FOR I = 1 TO  LEN (A$)
100  B$ =  MID$ (A$,I,1)
105  F = 0
110  FOR J = 1 TO N: IF X$(J,1) = B$ THEN C$ = X$(J,2):F = 1
120  NEXT : IF F = 0 THEN 300
125  PRINT C$;"    ";
130  GOSUB 400
140  NEXT
150  PRINT : PRINT "MESSAGE COMPLETE"
160  END
300  PRINT : PRINT "*** SORRY, CHARACTER ";
310  PRINT B$;" IS NOT IN MY ALPHABET"
320  PRINT
330  PRINT "PLEASE TRY AGAIN."
340  F = 0: GOTO 50
400  REM  *** OUTPUT C$ ***
410  FOR K = 1 TO  LEN (C$)
420  D$ =  MID$ (C$,K,1)
430  IF D$ = "-" THEN  GOSUB 700: GOTO 470
440  IF D$ = "." THEN  GOSUB 800: GOTO 470
450  IF D$ <  > " " THEN 600
460  GOSUB 900
470  FOR T = 1 TO 60: NEXT T
480  NEXT K
490  FOR T = 1 TO 200: NEXT T
500  RETURN
600  PRINT : PRINT ":CODING ERROR ";C$
610  STOP
700  REM  ***DASH ***
710  FOR T = 1 TO 40
720  SOUND =  PEEK ( - 16336) * 2
730  NEXT T
740  RETURN
800  REM  *** DOT ***
810  FOR T = 1 TO 10
820  SOUND =  PEEK ( - 16336) * 2
830  NEXT T
840  RETURN
900  REM  *** DELAY ***
910  FOR T = 1 TO 40
920  NEXT T: RETURN
1000 REM  *** THE MORSE CODE ***
1010 DATA  36
1020 DATA  A,.-,B,-...,C,-.-.,D,-..
1030 DATA  E,.,F,..-.,G,--.,H,....
1040 DATA  I,..,J,.---,K,-.-,L,.-..
1050 DATA  M,--,N,-.,O,---,P,.--.
1060 DATA  Q,--.-,R,.-.,S,...,T,-,U,..-
1070 DATA  V,...-,W,.--,X,-..-,Y,-.--,Z,--..
1080 DATA  0,-----,1,.----,2,..---,3,...--
1090 DATA  4,....-,5,.....,6,-....,7,--...
1100 DATA  8,---..,9,----.
1110 DATA  ;,.-.-..
1120 DATA  .,......
```

SECTION 3

The Design Process

Situation and brief

Flow chart drawing
A design process by which a technological project can be solved was outlined on pages 4 and 5. Ten examples given in pages 10 to 44 show how this design process can be applied to the solution of school technology projects. A flow chart outlining the design process is included on this page. This is similar to that already shown on page 4, except for the omission of the stages in the analysis and investigation. In this section of the book some of the stages in the design process as they apply to school technology are now given further attention.

Situation
The process of designing is the attempt to find solutions to *situations* arising from the needs of people. Such needs can arise from the desire of people to perform some task or to undertake some form of action. A solution may be required when the situation is that an existing design is unsatisfactory or where there is a need to improve or modify an existing design. Another type of situation which can arise in schools and colleges is when a teacher or lecturer arranges a situation in order that pupils and students can have practice in designing.

Because designing involves attempts at finding solutions to situations, the situation should be clearly stated in writing before the problems involved in a design are investigated. Then, when the design has been completed, the designer can refer back to the written statement of the situation and check whether the completed design meets the requirements of the statement.

Design brief
A design brief should state quite clearly what is to be designed in order to meet the requirements of the situation for which the design is being produced. The purpose of a design brief is to select an appropriate statement as to how the needs of a situation can best be met.

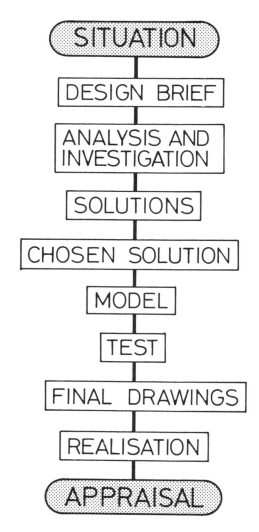

A flow chart outlining the design process

The design brief should always be written down in order to check, at a later stage, whether the solutions to a design problem are meeting the requirements of the design brief. A design brief could well be changed or amended during the designing process, if it is found that the brief cannot be met. Thus, whereas the situation for which one is designing cannot be changed, the design brief arising from the situation can be amended if necessary.

Some examples of situations and design briefs

Situation
An elderly disabled man finds great difficulty in climbing the stairs in his house in order to reach his bedroom.
Design brief
Design a method which is suitable for carrying the disabled man up the stairs.

Situation
"In my kitchen there is always a smell of stale cooking."
Design brief
Design an air conditioning system which will keep the air in the kitchen fresh.

Situation
"I would like to photograph my house and garden from above."
Design brief
Design a device which can be carried above the house by a balloon and which will carry a camera and, by remote control, take a photograph of the house and garden when it reaches a height of 100 metres.

A variety of briefs for the same situation
A complex situation such as the following, can give rise to a wide variety of design briefs.

Situation
Each day 10 000 items of stock from a factory store have to be transported to different points along a production line at different times of the day. Production must not be stopped while the items are delivered at the points on the production line.
A general brief to this situation
Design a method by which the stock items can be transported from the store to the production line.

It is more likely, however, because of a variety of restrictions of facilities, space, costs etc. that a more specific design brief will arise from this situation. Some examples are as follows:
Specific design briefs
1. Design a vehicle to be driven from its own electric battery power supply which can be used to transport the items of stock from the store to the production line.
2. Design a conveyer belt system on to which the items of stock can be placed in the store and conveyed to the production line.
3. Design a hand truck which can be controlled by one man and used for carrying the items of stock from the store to the production line.

4. Design an overhead carrier system on which items can be hung in the store and conveyed to the production line.

Rules
1. Always write down a clear statement of the situation.
2. Always write down a clear statement of the intended design brief arising from a given situation.
3. The situation normally cannot be changed or amended.
4. A design brief may be amended or, if necessary, changed, if solutions to a stated design brief are found to be difficult or even impossible to solve.

Exercises
A list of possible technology projects is given below. Write a statement of a possible situation and a statement of a possible design brief which you consider suitable for each of the projects.
1. A control system for a model powered aeroplane.
2. An automatic litter picker.
3. Shelf identification of materials kept in a store.
4. A turntable for a model railway system.
5. The production of methane gas from farm waste.
6. An automatic barrier raising device to control road access to a building.
7. A disposal unit for farm effluent.
8. An automatic feed system for farm animals—e.g. pigs, cattle, poultry.
9. A drinks dispensing system.
10. A model automatic car parking system.
11. Automatic traffic lights system.
12. A robot capable of following a sound.

Note
The design process should therefore commence with two written statements.
1. The *situation*.
2. A *design brief* which identifies a method suitable for solving the problems arising from the situation.

Analysis and investigation

The flow chart on this page repeats, in a different form, the Analysis and Investigation part of the Design Process chart on page 4. Ten headings under which the problems associated with designing for a technology project can be analysed and investigated are shown in the sequence illustrated in this flow chart.

In the sense used here—that of designing for technology—the word *analysis* means that a thorough examination of the various elements involved in seeking a good solution to a design brief should take place. Each of the elements needs to be *investigated* in order to discover the details which can affect the finding of a good solution by which a design brief can be said to be satisfied.

Ten headings under which the problems associated with designing for a technology project can be analysed and investigated are shown in the flow chart. While these are given in what would appear to be a logical order, there is no reason why the investigation must follow the given sequence. For example, in some project design work, it may be thought necessary to investigate the strength of different groups of materials, before it is possible to select the best material for the particular function expected of it; the fittings to be used may have to be selected at the same time as the best methods of jointing parts of the design to each other are being sought; safety in some technology projects may be of paramount importance, in which case safety might have to take priority in the list of details to be considered when designing for such projects. The ten headings are given only as a guide and the sequence of the analysis and investigation may vary between different types of project and according to the methods of the designer. Some of the headings may not apply to some projects, in which case they can be ignored.

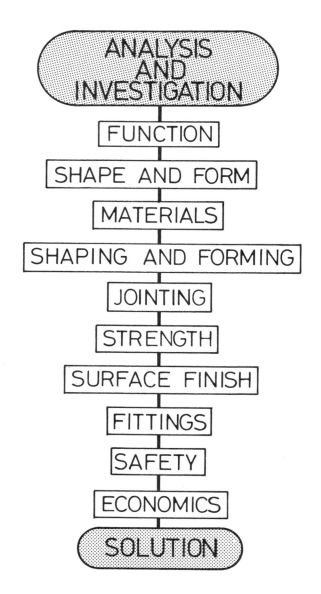

Notes and drawings
No matter which sequence is selected when investigating the factors involved in designing for a technology project, one feature must be stressed. The keeping of a good record of the details showing the progress of an investigation is of the greatest importance. In this book the use of good drawings is shown to be of equal importance as the writing of good notes. A record of the details in the designing process must be made. All ideas and suggestions should be noted, no matter how absurd or trivial they may seem at the time. Even ideas which seem completely wrong may give rise to other suggestions being considered. Constant reference to the notes and drawings will be found to be necessary as the search for good solutions proceeds. Without a record of good notes made during an investigation, many details of difficulties encountered and overcome might well be lost.

Design folio covers
A good set of covers—a folder or a folio—in which all notes and drawings referring to a particular project can be stored, is recommended. A large envelope (a manilla card envelope) of A4 or A3 size; a good ring binder of A4 size; a set of home-made cardboard covers of A4 or A3 size; a good folio cover of A3 or even A2 size—home-made or purchased. Any such set of covers can be chosen. A well designed cover for the folder or folio is also recommended. A well designed folio will give the owner a feeling of pride in his or her design work and will lead to better designing. The design of the covers in which the notes and graphics are kept should be regarded as an essential feature of the design process.

Function

A completed design, intended to meet the requirements of any situation and its attendant design brief, must *function* properly. If a design does not function well, then the attempts at solving the problems involved in meeting a design brief have failed. In school technology, some projects may have the achievement of purely functional design, as their main or even their sole aim. Even when function is not the primary aim of a technology project, the completed design should operate in such a manner as to properly perform the tasks for which it has been designed. No matter how beautiful or elegant a solution to a design appears to be, if it does not function well, it may just as well not have been produced.

Before commencing an analysis and investigation into the requirements of a design brief, it is as well to consider and then write down a list of what you consider to be the true functions of the design you hope to achieve.

1. For what purpose or purposes is the design being made? What is its function, or what are its functions.
2. What should the design achieve when it has been made?

Very often the appearance of a design based in school technology is not of great importance and correct functioning plays a much more important role than does final appearance. This does not mean that appearance should be ignored. Many purely functional designs are in themselves of good appearance, although relying completely on correct functioning to solve the aesthetic value of a design does not necessarily achieve good results.

Exercises
Four design briefs suitable for school technology projects are given below. List the functions you consider appropriate to each of the completed projects.
1. Design a wind tunnel in which the high velocity part is 300 mm square in cross section.
2. Design an apparatus which could be used for testing the glue lines of pieces of wood held together with different adhesives.
3. Design a device to monitor gas passing between two storage tanks in a chemical works.
4. Design an automatic feeding device for fitting into a pigeon loft for dispensing seed to pigeons upon their return home after a flight.

Shape and form

In the paragraphs dealing with Function, it was suggested that the shape and form of a design for a technology project will be influenced by purely technological considerations. This suggestion means that the appearance of a completed design may have to be sacrificed in order that it can function as required by the design brief. Despite this, the shape and form of a design for a technology project should be as carefully considered as all other aspects in the design process. If a completed design functions well and is also of good appearance and is aesthetically pleasing, then its design brief has been satisfied more completely than if only technological function has been considered.

The shape and form of a design for a technology project may depend upon such features as:

Ergonomics
Is the design of a shape and form suitable for it to be used with ease and comfort by those for whom it has been designed? This *ergonomics* problem applies not only to the overall design but also to its parts. For example, are controls in a suitable position for easy use? Are the controls of a suitable size? Is their colour important? Should they carry symbols to clearly show their function? Are they safe to use?

Weight
Is the design too heavy or too light in weight? If it is too heavy to be used with ease, can its shape and form be amended without affecting the proper functioning of the design? Might the functioning of a design be improved by increasing or decreasing its weight? Is the design to be stationary or is it to be mobile—will it be operated from a remote control or will it be operated where it stands? Must the design be buoyant in water, or float in air?

Wind and water resistance
Must the design be aerodynamically designed so that its outer shape is such that it can move through the air freely? Does it require to move through water and be designed so that water resistance to its movement is reduced to a minimum?

Materials
Safety aspects may influence shape and form because the materials which have to be used may need to be electrical conductors or insulators. The materials may be such that they may have to be air-tight or water-resistant; they may have to withstand intense abrasion or be sufficiently smooth to reduce friction. Such features, among others, can well affect shape and form in that the methods of craft-working available may not be able to achieve a required shape and form if the materials to be used are dictated by the functional demands of the design.

Centre of gravity
The positions of working parts, of features such as motors and other parts which are heavy in relation to the overall design, may affect the centre of gravity of a design. If the design is one which involves flight in air, movement on or in water, or even movement along a hard surface, the centre of gravity of the whole design may well have to be carefully considered. The positions of heavy parts is important in such designs and can affect their overall shape and form.

Exercise
1. Write down a list of other technological features which you think may affect the shape and form of a design. Some examples can be taken from the ten projects described in pages 10 to 44.

Materials

The materials for constructing the designs resulting from school technology projects will often be selected from the same groups of materials from which designs based on craftworking are produced. These materials are commonly found in the stores of Design, Craft and Technology departments in schools. Such materials cover a wide range including woods, metals, plastics and other materials.

Woods—Various softwood and hardwood species; a variety of laminated boards such as plywoods, blockboards, laminboard, hardboards, chipboards; veneers; dowels.

Metals—Ferrous metals such as mild steels, tool steels, sheet mild steel, tinplate bars, rods, wires, special turning steels; non-ferrous metals such as copper, brass, aluminium, zinc in bar, rod, sheet and extruded forms.

Plastics—'Solid' plastics in sheet, rod and extruded forms; resins for making glass reinforced plastic (GRP) mouldings together with the associated glass fibre materials; resins for casting and straight mouldings; foams of various types.

Other materials—Materials such as clays, cements, plasters, glass, upholstery materials; nails, screws, nuts, bolts; a variety of adhesives for wood, metal and plastics; a variety of soldering, brazing and welding materials together with their respective fluxes.

Materials for technology projects

The requirements of meeting the demands of designing for technology projects add yet more materials to the above four groups. Some of the materials in this extra group are described here. It should be noted, however, that a list of all materials used in technology projects would prove to be too extensive for inclusion in a book of this nature.

Meccano

Meccano construction kits are widely used, not only for the making of models, but also for building designs for projects for both educational and industrial applications. Photographs of items from the Meccano range, together with an example of a Meccano built school project and an industrial project built from Meccano are given on this page. Meccano constructions are firm yet flexible, strong but light in weight and are suitable for both temporary and permanent use.

All-metal structures containing mechanisms dependent upon—gears and gear boxes; pulleys and pulley systems; axles and wheels; steering systems; mechanical link systems; link mechanisms and the like; can be made from Meccano units. Meccano built mechanisms can be driven by small electric Meccano motors which are dependent upon a 12 volt direct current power supply—supplied from battery or power pack. The motors themselves are available with integral gear systems which can provide a variety of spindle speeds.

Meccano structures are held together with nuts and bolts. If Meccano nuts and bolts are used, it is advisable to have a set of $\frac{5}{32}$ in British Standard Whitworth (BSW) taps and dies at hand when using Meccano for project work. If this is not thought to be advisable, it may be considered preferable to purchase 4 mm nuts and bolts for building Meccano structures. This allows the use of 4 mm diameter silver steel rod for axles and also when strips of rod which need a screw thread are required as part of a structure.

Some items of Meccano construction kit pieces

An example of a structure made from Meccano. A model vehicle, part of a light-seeking device. Made from Meccano strips, angles, plates, wheels and gears with electric motors

An industrial use of Meccano. This 'Mains Buggy' was designed to fault find along a distribution gas mains. The 'Buggy' carries a television camera mounted on a structure made from Meccano items running on Meccano wheels. Photograph by courtesy of the Central Laboratories of SEGAS.

Structures made from Meccano, wood and plastic sheet material can be constructed. When making up gear boxes, sides made from acrylic sheet with accurately drilled holes to receive the gear axles, may be preferable to the sheet metal gear box cases made from Meccano.

A difficulty when using Meccano items for technology projects lies in the storage of the multitude of different sizes and shapes of the pieces. This storage problem can be overcome by the use of plastic trays in which compartments have been fitted. Vacuum formed trays/compartments made from thermoplastic sheet could be purpose-made for storing these pieces.

Hybridex

An aluminium extrusion of the sectional shape shown in Drawing 1 is available under the trade name of Hybridex. The extrusion is purchased in standard lengths and cut to size as required. A special jig can be purchased to assist in sawing the required lengths with ends at exactly 90° to allow accurate jointing. Hybridex is a material suitable for making model structures intended to be permanent. Drawings 2 and 3 indicate how the cut lengths of Hybridex are jointed to form rigid structures.

Danum–Trent electrical components mountings

A system for the mounting of electrical and electronics components for building up circuits was developed at the Danum Grammar School, Doncaster, under the auspices of the National Centre for School Technology (NCST), Trent Polytechnic.

The system is based on two different plastic extrusions, one of which is a press, or a sliding, fit into the second. The first of these extrusions can be drilled with holes into which electrical components and sockets for 4 mm plugs can be fitted. The electrical or electronics components can be mounted either above or below the upper surface of the lengths of plastic. The mounted components can be clipped into lengths of the second extrusion for mounting on work under development. Amendments to a circuit can be speedily made by changing components or by changing connections. 4 mm stackable plugs with leads are used to form conductors connecting the components to each other. Examples of the use of Danum-Trent mountings have already been given earlier in this book. Two photographs on this page show other examples.

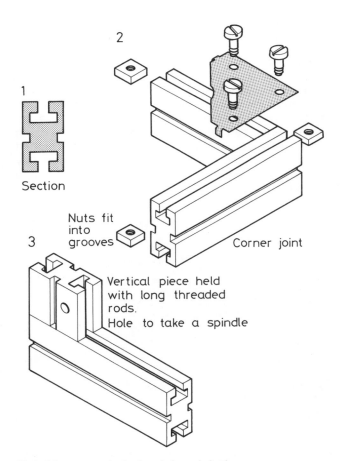

Hybridex extruded aluminium jointing

Danum-Trent circuit used with a model vehicle constructed from Meccano. Power obtained from a NCST power pack

Two Danum-Trent component mounts

Fischertechnik systems

Among other construction systems which have been developed and are suitable for technology in schools are the Fischertechnik systems. This constructional system comprises a whole range of precision manufactured parts made from a tough plastics material. These parts can be interlocked to each other to form firm, strong structures. Gear and pulley systems and wheels and axles form part of the kits. Electrical and electronics kits are also made in Fischertechnik for use either for the building of circuits or for use with structures made from the construction system.

Strip material
Wood

Wood strips such as the offcuts obtained when timber is sawn to size on a circular saw, or purpose-sawn strips cut to size for use in technology, make excellent constructional material for some project work. Suitable strip sectional sizes may be 6 mm square; 12 mm by 6 mm; 9 mm square; 9 mm by 6 mm among others. Wood strips can be jointed by gluing with PVA glue and taping with masking tape or, if holes are bored to receive them, screws and light nails (veneer and panel pins) make suitable joints, providing the joints are also glued. Movable joints to form link mechanisms can also be made if hardwood strips are used, but care must be taken to ensure that holes are accurately bored to receive the pins on which the links can pivot. Quite large and strong constructions are possible using wood strip materials of small sectional dimensions. Very large constructions are possible with larger strips—say of 25 mm square, or of 20 mm by 10 mm and other such larger sizes.

Care is required when constructing from wood to avoid splitting at joints. Thus glued and taped joints may be preferable to nailed ones.
Metal

Strips of mild steel and strip aluminium are excellent constructional materials for technology. Strips of sectional sizes of 12 mm by 3 mm, or 9 mm by 2 mm, can be bent with ease, can be drilled accurately and can be rivetted and bolted to form strong constructions. Mild steel can also be brazed or welded but aluminium needs special fluxes and brazing material before it can be hard soldered easily. Metal strip material can be regarded as being preferable to wood when link mechanisms or projects which involve moving parts are being constructed. This is because metal can be drilled with greater accuracy than can wood and is much harder wearing.
Plastics

Strips of plastics material such as acrylic (e.g. Perspex) cut from 3 mm thick sheets to widths of say 6 mm, 9 mm or 12 mm, provide excellent constructional materials for technology project work. Such strips can be built into strong, firm yet flexible constructions, if the methods of construction are well designed. Strips of acrylic can be glued to form strong joints if the correct adhesives are used and can be bolted with nuts and

An example of the use of fibre glass for reinforcing a box construction made from plywood. The photograph shows a bonnet under construction for a 'jeep' like vehicle

bolts. Acrylic strips can also be rivetted with aluminium rivets providing great care is taken to avoid splitting when the rivets are hammered home. Holes can be bored in acrylic materials with ease and accuracy, thus moving parts can be constructed from this material. Acrylic strips can be heat bent at temperatures from the boiling point of water to 130° C.

Sheet materials
Wood

Laminated boards such as plywoods, hardboards and chipboards are valuable materials for use in technology project work. They are all relatively cheap materials, are easy to work, can be jointed to form structures using simple methods and are strong if used in well-designed constructions. They also have the advantage that they are insulators against electrical current at the low voltages used in school technology work.

A gear 'box' made from acrylic sheet mounted on strips of softwood

Metal

Sheet mild steel and tinplate are materials which can be cut to quite intricate shapes and can be bent with ease. Mild steel can be brazed and welded to form extremely strong, permanent joints, but tinplate can only be soft soldered. Mild steel sheet can be drilled with accuracy to provide holes for bolting or for rivetting or to receive axles or spindles of moving parts. Tinplate can have holes bored in it, but the degree of accuracy obtained not only in siting holes but in the actual hole diameter is not as good as for mild steel sheet.

Aluminium sheet is a good sheet material for technology projects. It also takes hole boring with accuracy for bolting or rivetting. Aluminium sheet will bend easily, but complicated bent shapes in aluminium involve occasional softening of the sheet by annealing because aluminium work-hardens as bending proceeds. As with strip aluminium, special fluxes and solders are necessary if this metal is to be heat jointed. All these metals conduct electricity even at the low voltages used in school technology projects.

Plastics

Transparent clear sheet acrylic in thicknesses of 2 mm or 3 mm is an excellent material for those projects in which internal parts need to be clearly seen. Any sheet plastic materials, whether transparent or opaque, are useful materials for technology. They can be drilled, sawn, filed to shape, polished and bent (heat bent) using elementary craft skills. All these plastic materials are excellent electrical insulators.

Glass reinforced plastic

Mouldings made from glass fibre and polyester resins can be used to advantage in some project work. The making of GRP mouldings involves relatively simple craft skills, but can be somewhat 'messy' unless a workshop is equipped to deal with such work. When making lightweight structures of box form from plywoods or hardboards, the corner joints of such structures can be considerably strengthened by the addition of two or three layers of glass fibre 'laid on' with polyester resin. Joint lines of such structures can be jointed specifically with fibre glass and resin. First bore matching holes along both of the edges to be joined. Then 'sew' the edges together by looping lengths of copper wire through the holes and tighten by twisting with pliers. Finally 'lay-up' the joint with bands of glass fibre and polyester resin. Such joints are extremely strong.

Silver steel

Silver steel is a high quality carbon steel which can be hardened and tempered in the same way as carbon tool steels. Its surfaces are ground to very fine degrees of dimensional accuracy—hence the term 'silver'. Because of this high degree of dimensional accuracy, silver steel rods make extremely good axles for wheels or spindles for other working parts. In particular, silver steel rod of 4 mm diameter is compatible with Meccano equipment. However a whole range of sizes of silver steel can be obtained.

Soft iron wire

Soft iron wire of 16 SWG diameter can have many uses in technology. In particular it can be used for the temporary cramping together of parts which require to be glued, soldered or brazed. Although soft and hence easily twisted, its strength values are good.

Mild steel rod

Another useful metal material for project work is mild steel rod of 3 mm diameter. This thin rod can be easily bent to shape, can be brazed or welded and is thus easy to join. It is valuable when seeking strengthening rods or ties in lightweight metal constructions.

Scrap materials

Second-hand materials given as 'gifts' or bought cheaply or dismantled from items no longer of use, should always be examined with some care. The most useful of such items are those which have electrical, electronic or mechanical parts.

Most of what is obtained second-hand can be thrown away, but if care is taken in dismantling the material not only will a great deal be learned about industrial constructional methods, but many items which can be used for projects will be found. Anything unusual should be kept and a simple storage system devised for the material so saved. Junk in junk boxes remains junk. Properly dismantled and stored such items as switches, resistors, gears, pulleys, nuts and bolts, spindles and so on, can help in the solving of many problems which arise when attempting solutions to a technology project. Such items as transistors however and other semi-conductors are not worth de-soldering and storing.

An important safety aspect is to NEVER attempt connecting such second-hand devices to a power source just to 'see how it works'. First, it will probably not work and, secondly, it may be a source of danger if, for example, very large capacitors suddenly discharge after being put back in store because the device is not functioning.

Other materials

Solutions to technological problems can often give rise to requests for a wide variety of different materials. In school technology requests such as 'Have we any cement?'; 'Where can I find an inner tube?'; 'Is there a retort stand in the workshop?' are often heard. Among other items which have been requested are sugar; salt; paper clips, vinyl rainwater guttering; vinyl pipe; table tennis balls; marbles; elastic bands. The list can appear endless, but re-inforces the point that, when seeking solutions to a technology problem, it is advisable to be on the look-out for any materials which can assist in the solving of the problem.

Methods of construction

This part of the book has been included in order to show methods of craft constructions which may be necessary when constructing, assembling and making projects for school technology.

Temporary electrical and electronics circuit constructions

A variety of methods of constructing circuits of a temporary nature has been shown in photographs on pages 72 to 94. Such methods are of value not only when experimenting in order to understand how components function in circuits. They can also be employed to test circuits designed to perform specific functions as part of a project. The methods already shown include:

1. School-made wood boxes with 4 mm sockets connected to individual components.
2. 'Danum' sets with sockets and components mounted in extruded plastic moulding. Several methods of using the extrusions are shown.
3. Using crocodile clips to hold components.
4. Using S-DeC platforms.

Yet another method is shown in a photograph on this page—using Fischertechnik units. Other manufacturers make different units of this type.

The making of more permanent circuits

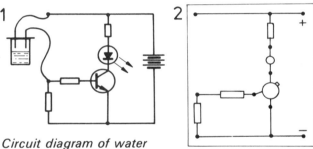

Circuit diagram of water level indicator

Full size plan of the circuit

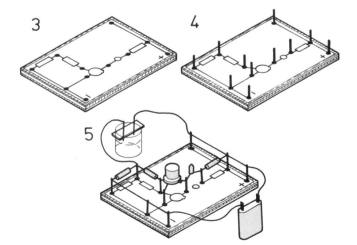

Soldering to pins

To construct the circuit shown in Drawing 1.

1. Draw a new circuit drawing but showing the outline of each component—Drawing 2.
2. Glue the drawing to a piece of plywood or other suitable board—Drawing 3.
3. Hammer brass pins partly into the board at the positions shown by the dots of Drawing 2—Drawing 4.
4. Solder the components to the board, using tinned wires soldered between and to the brass pins.

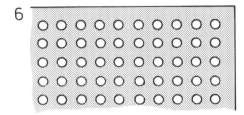

Matrix board (twice full size)

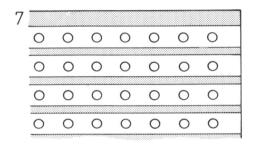

Stripboard (twice full size)

Matrix board

Matrix board is made from synthetic resin bonded paper (phenol-formaldehyde). It can be purchased in small panels of about 150 mm by 100 mm. It is pierced with small holes of 1.3 mm diameter at 2.5 mm centres. The holes will accept the connectors of many components. Pins can be purchased which can be pressed tightly into the holes, allowing circuit assembly as for the method shown on page 112. Wires are simply soldered to the pins. Circuits can also be assembled as shown— Drawing 8.

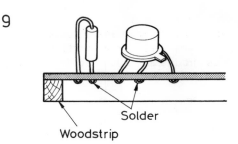

9

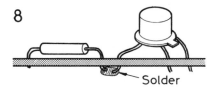

8

Solder

Woodstrip

Veroboard or stripboard

This is also made from synthetic resin bonded paper. Copper tracks are bonded to one side of the board and 1.3 mm holes at 3.8 mm centres are drilled through the board and tracks. Components are mounted on the plain side of the board and their connectors are soldered to the copper tracks on the other side. It may be necessary to 'break' the copper track. This is done by either cutting out a strip of the copper with a knife or using a purpose-made tool which drills away the track. Building circuits on stripboards needs careful planning and it is advisable to plan a circuit on grid paper before attempting to construct it on stripboard.

Because the underside of matrix board and stripboards become uneven it is advisable to mount the boards on strips of wood. Allowances for the widths of these wood strips will need to be made when designing a project into which circuits on matrix or stripboards are to be mounted.

Printed circuit boards

Printed circuit boards can be etched to one's own designs. Plain copper-clad boards for this purpose can be purchased. The circuit can be drawn on to the copper with an etch-resist pen or etch-resist transfers can be purchased. Etching solutions which can be used to dissolve out the unwanted copper are sodium hydroxide and ferric chloride. It must be noted that solutions of these chemicals must be treated with care; there is always a safety risk if they are used unwisely.
1. Draw or trace the circuit on to the copper.
2. Go over the circuit with an etch-resist ink or with transfers.
3. Place the board in a ferric chloride solution—in a plastic tray—face down for about 20 minutes, 250 g of ferric chloride per $\frac{1}{2}$ litre. Wear rubber gloves, use tongs and wear goggles.
4. Remove board, wash thoroughly in clean water and then clean with pcb eraser—which will clean and degrease the copper.
5. Drill holes as necessary to receive the connectors of the components.

Heat-sink soldering

All semi-conductors must be soldered using some form of heat-sink method (see later pages). When including integrated circuits in a circuit always use the special sockets designed to receive them.

Another type of board on which circuits can be temporarily constructed

Soldering connections to electrical components

Methods of making permanent soldered connections between cables and components in electrical and electronics circuits are illustrated on this page.

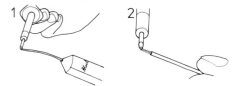

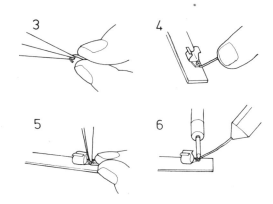

General method of making soldered joint
1. Pick up solder with electrical soldering iron
2. Bare cable end. Tin bared end with solder
3. Bend the tinned cable end with pliers
4. Loop bent end into hole in component
5. Close the bent loop with pliers 6. Apply solder to the joint to make it permanent

'Heat-sink' soldering with pliers. Heat of soldering absorbed by pliers. Pliers held on to transistor connection with several elastic bands looped over handles

'Heat-sink' soldering using a blob of 'Blu-tack' to absorb the heat and so prevent the heat from a soldering iron from damaging the electrical component

'Heat-sink' soldering using potato pieces
1. Electric soldering iron, flux and solder 2. Clean soldering bit by filing 3. Dip heated soldering bit in flux 4. Cut a piece from a potato 5. Push connector of transistor through potato piece 6. Apply solder between cable and connector

Some metalworking processes

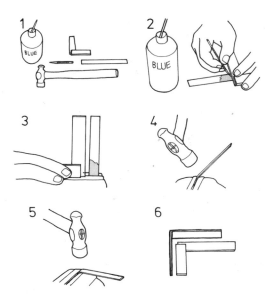

Bending strip metal
1. 'Blue', scriber, try square, hammer, metal strip
2. Apply 'blue', square scribed line at bend
3. Place strip in vice. Check squareness to vice jaws
4. Commence hammering bend 5. Complete hammering bend 6. Check completed bent strip for squareness

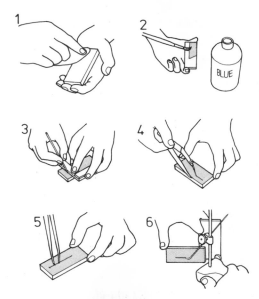

Marking out metal with scriber
1. Clean grease and dirt from surface 2. Apply 'blue' with a brush 3. Marking with try square and scriber 4. Marking line parallel to an edge with odd-leg dividers 5. Marking equal spaces along a line with dividers 6. Marking line parallel to an edge with scribing block

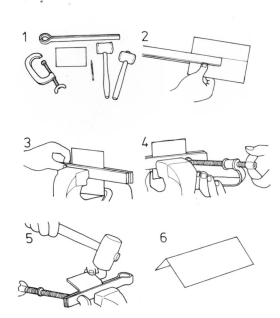

Bending sheet metal
1. G clamp, folding bars, scriber, mallets, sheet metal 2. Mark line of bend. Place sheet in folding bars 3. Place folding bars in vice 4. Tighten end of folding bars with G clamp 5. Use mallet to bend the sheet metal 6. Completed corner bend

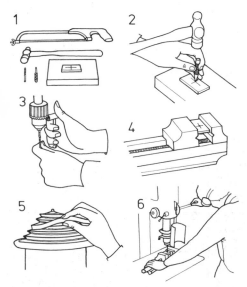

Drilling a hole in metal
1. Hacksaw, hammer, scriber, centre punch, metal on bench block 2. Mark hole position with centre punch 3. Place drill in chuck of drilling machine. Tighten 4. Place metal in machine vice 5. Adjust belt to give a suitable chuck speed 6. Drill hole. Note—drill guard should be down, not up as shown. The guard is shown in the up position in the drawing for the sake of clarity. Note also that goggles should be worn to protect the eyes

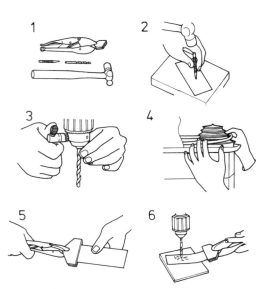

Drilling holes in thin sheet metal
1. A pair of 'Mole' grip pliers, centre punch, drill, hammer. 'Prima clamps' can also be used for this purpose   2. Mark hole position with centre punch   3. Fit drill in drilling machine chuck and tighten   4. Adjust speed of drilling machine   5. Place sheet metal in 'Mole' grips   6. Drill the hole

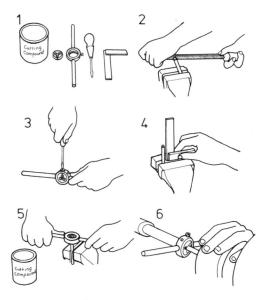

Cutting an external screw thread
1. Cutting compound, screw die, die stock, screwdriver, try square   2. File chamfer on rod on which thread is to be cut   3. Place die in stock and tighten   4. Check that rod is vertical in vice   5. Cut the thread—half-turns clockwise followed by three-quarter turns anti-clockwise. Use cutting compound to ease the cutting action   6. The thread can also be cut on a lathe, turning the lathe chuck by hand

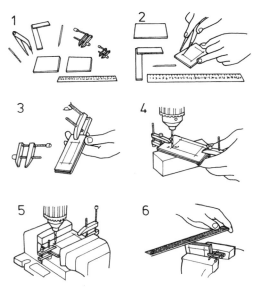

Boring matching holes in sheet metal
1. Drill, odd-leg dividers, scriber, 2 engineers cramps, ruler, the two pieces of metal   2. Mark hole positions   3. Cramp the two pieces together   4. Drill the holes, either on a block of wood OR   5. In a machine vice   6. File the two pieces of metal to size while still in the cramps

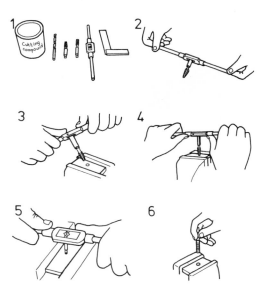

Cutting an internal screw thread
1. Cutting compound, drill, taper tap, plug tap, tap stock, try square   2. Place the taper tap in the tap stock   3. Insert tap in hole bored in metal   4. Check that tap is square to metal   5. Cut screw with clockwise half-turns each followed by anti-clockwise three-quarter turns to break swarf. Apply cutting compound to ease the cutting action   6. Check that screwed part fits the tapped hole

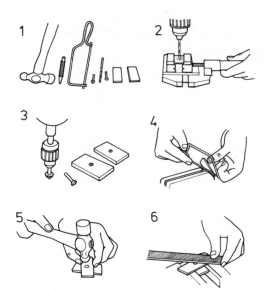

Rivetting
1. Hammer, centre punch, 'junior' hacksaw, countersink, drill, rivet, two pieces of metal to be joined
2. Centre punch and bore holes of rivet diameter
3. Countersink each hole   4. Put rivet through holes; saw off waste rivet   5. Hammer excess rivet into countersunk hole   6. File off rivet level with surface of metal

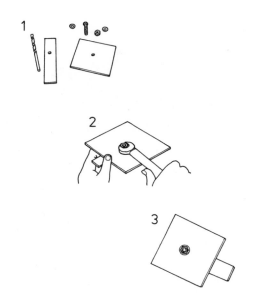

Bolting pieces together
1. Drill, pieces of material being joined, bolt, nut, two washers   2. Place bolt through holes with washer each side. Tighten nut with spanner   3. The completed join

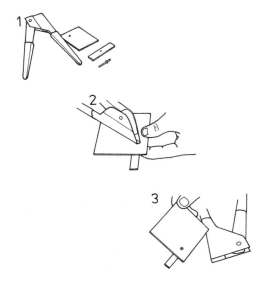

'POP' rivetting
1. POP rivet gun, sheets of metal with holes, POP rivet   2. Place rivet in gun, insert rivet through holes, apply pressure to gun handles until excess rivet breaks away   3. Completed rivetted pieces
Note—POP is a registered trade mark for fastenings produced by Tucker Fastenings Limited

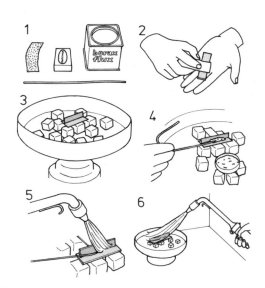

Brazing
1. Emery cloth, matches, flux, brazing rod   2. Clean surfaces to be brazed with emery cloth   3. Place parts being brazed in brazing hearth   4. Apply flux   5. Heat to cherry red and melt brazing rod into join   6. Continue heating until molten brass runs right through joint

Oxy-acetylene welding
1. Oxygen cylinder, acetylene cylinder, torch, goggles. Note—must be welding goggles 2. Turn on acetylene and ignite 3. Introduce oxygen to produce a welding flame 4. Tack the ends of the piece being welded 5. Use welding rod to weld the joint 6. Completed weld. Safety gas welding equipment should only be used under trained, qualified instructors or teachers.

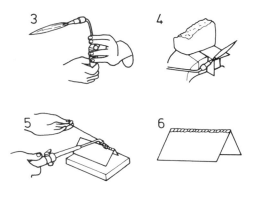

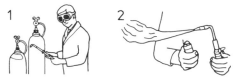

Some joints in pneumatics and hydraulics circuits

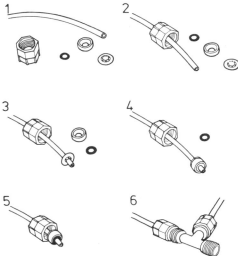

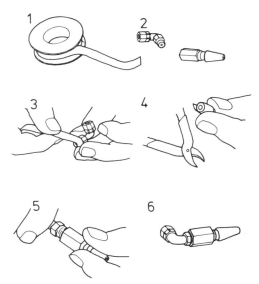

'O' ring connections in pneumatic circuits
1. Pneumatic plastic piping, nut, cup washer, O ring, serrated washer 2. Push nut over piping 3. Push serrated washer over piping 4. Push cup washer over piping 5. Push O ring over piping 6. Tighten joints

PTFE taped joint
1. Roll of polytetrafluoroethylene tape 2. Parts to be water-tight jointed 3. Wrap screw thread with PTFE tape 4. Cut excess tape off 5. Screw parts together 6. Completed join

Some woodworking processes

Making a wooden box
1. Hammer, box sides, pva glue, nail punch   2. Place nails dovetail fashion in ends of long sides   3. Drive home nails in one of the box corners   4. Nail another box corner   5. Complete nailing box sides   6. Add bottom, nail in place, punch all nail heads below surface

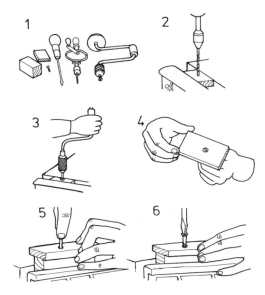

Making a screwed joint
1. Pieces being jointed, wood screw, screwdriver, handdrill, carpenters brace with countersink   2. Bore hole to take screw shank in upper piece of wood   3. Countersink hole   4. Test the screw head in its countersink   5. Place pieces together. Bore pilot hole in lower piece with bradawl   6. Screw joint together

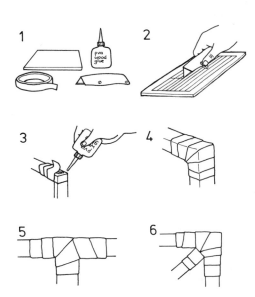

Temporary joints in wood strips
1. Cutting board, masking tape, pva glue, cutting knife   2. Cut masking tape into lengths on cutting board   3. Glue joint, hold together with masking tape   4. Completed corner joint   5. Completed T joint   6. Completed corner joint, braced with a diagonal

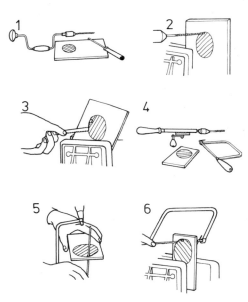

Cutting large holes in sheet wood
1. Sheet wood thicker than three-ply, brace and 12 mm bit, pad saw or compass saw   2. Bore a hole at edge of circle with the brace and bit   3. Insert saw through hole and cut the hole to shape   4. Sheet of three-ply or similar thickness sheet wood, hand drill and drill, coping saw   5. Bore hole near circle edge and insert coping saw blade.   6. Tighten coping saw blade and saw hole to shape

Some plastics working processes

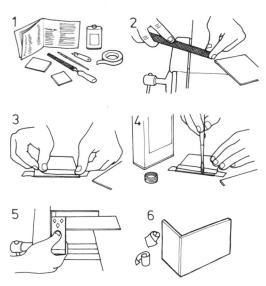

Edge jointing sheet plastics material
1. Adhesives directory, pieces to be jointed, file, brush, knife, masking tape 2. File edges to be jointed straight and square 3. Apply masking tape to prevent adhesive being brushed on to parts not being jointed 4. Brush on adhesive 5. Place jointed pieces in vice, test for squareness, leave until adhesive sets 6. Completed joint, masking tape removed

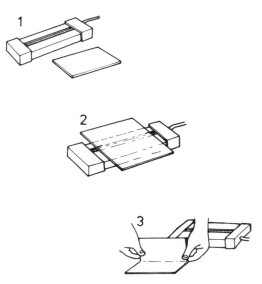

Bending sheet plastics on a strip heater
1. Strip heater, sheet to be bent 2. Place sheet on heater, switch on heater, wait until heated strip of plastic is flexible 3. Bend to shape. Hold until cool

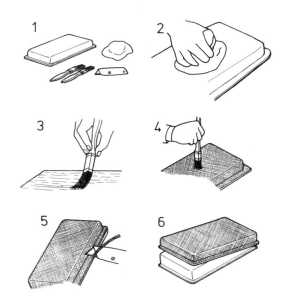

Making a fibre glass moulding
1. Mould, brushes, polishing cloth, knife 2. Apply release agent over mould surface with the polishing cloth 3. Apply a gel coat of polyester resin by brush 4. Apply chopped mat glass fibre and lay-up polystyrene resin with brush 5. At 'green stage' cut away excess fibre glass 6. When hardened, release moulding from its mould

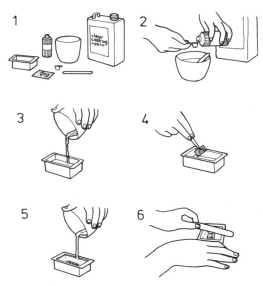

Encapsulation in polystyrene resin
1. Polythene moulding box, the object being encapsulated (an electronics circuit on matrix board), catalyst, mixing dish, plastic measure, clear casting resin, mixing stick 2. Add catalyst to the clear casting resin. Stir 3. Pour a layer of catalyst resin into the moulding box, allow to gel 4. Place article being encapsulated on to layer of resin 5. Top up with resin 6. Place polyester film on top. Smooth down with stick

Turning metal rod on a lathe

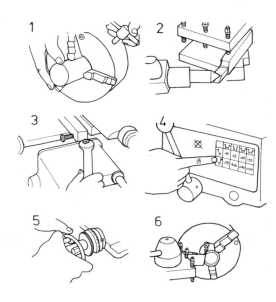

Facing off an end of a rod
1. Place rod in chuck. Tighten   2. Check that cutting tool is level with lathe centre   3. Tighten slide rest on to lathe bed   4. Set correct speed of lathe   5. Make cut by double hand movement of control on slide rest   6. Cut being taken off end

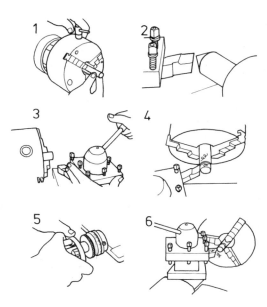

Turning a shoulder on end of a rod
1. Place rod in chuck. Tighten   2. Check that cutting tool is level with lathe centre   3. Check that cutting tool is at correct angle to end of rod   4. Make first cut   5. Put on a little more cut   6. Take second cut and so on until shoulder is formed

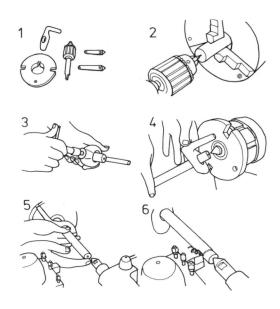

Turning between centres
1. Face plate, dog, centres, drill chuck   2. Drill holes each end with centre drill—centre drill held in tailstock   3. Tighten dog on to one end of rod   4. Place end with dog on driven centre   5. Place other end on dead centre. Tighten tailstock   6. Make a cut

Stitched and fibre glassed joints in sheet material

The photograph shows a series of joints being made with copper stitching and fibre glass to build a structure made from sheet plywood. The stages of producing such joints are as follows:
1. Stitch the parts together by passing lengths of copper wire through holes cut near the edges. Tighten each wire by twisting with pliers.
2. Work several layers of chopped strand glass fibre mat into the joint corners with lay-up polyester resin applied by brush.
3. Allow to harden

Strength–testing

Two areas of testing need to be considered in connection with project designing for school technology.
1. Testing the parts of a project as they are completed and testing the completed design.
2. Testing the materials to be used in a project.
A completed project is tested to see if it meets the design brief. The methods necessary for the test therefore depend on the function for which the design is intended. Tests should be carried out accordingly.

Materials testing

Obviously if the materials used in a project are unsatisfactory, there is a likelihood that the design will fail. It may therefore be necessary to test materials before including them in a design. A large number and variety of tests have been devised. As new materials come into use, new tests may themselves have to be designed. Such tests vary from the very simple ones that can be carried out in the classroom or at home to those which are more complicated, requiring measuring devices. More advanced tests may require a special laboratory or special test areas in an industrial concern.

Tests are comparative

An important detail which must be understood when testing materials is that all such tests are *comparative*. Materials tests are designed only to compare one material with others previously tested. Thus the tests can only determine how a material performs when measured against other materials similarly tested.

Simple classroom tests

Take pieces of a variety of different materials. Try to ensure they are all about the same size—preferably strips—woods, metals, plastics, cards, papers, wires, clays, strings, cottons, rubber materials and so on. Write down what happens when each strip is tested as follows:
State its colour.
Is its surface smooth, rough or otherwise?
Is it glossy, shiny, dull, matt, metallic or otherwise?
Is it heavy or light? Arrange the materials by weight.
Is it hard, soft, flexible, spongey or otherwise? Can you scratch it—with a finger nail, with a knife? Does it have a distinctive smell? Describe any such smell.
Does it have a distinctive taste? Describe any such taste (*Note*—some materials should *not* be tasted).
Can it be bent in your fingers?
If bent, does it stay bent?
Does it stretch by hand?
If stretched, does it go back to its original shape?
Is it springy?
Can it be shaped in your hands—is it malleable?
Does it fracture or split easily?
What tools will cut the material?
Is the material inflammable (*Note*—do not attempt setting fire to some materials).
Do you think the material is a good one to take a tension stress? A compression force? A flexing force? Will it sink in water?

Further school tests

By experiment in a science laboratory the following properties of materials may be examined and assessed:
Relative density—grams per cubic centimetre g/cm^3.
Melting point—in degrees Celsius $^\circ C$.
Freezing point—in degrees Celsius $^\circ C$.
Thermal expansion—coefficient of linear expansion.
Conductivity—of heat and/or of electricity.
Water resistance.
Resistance to corrosion.

British Standard tests for woods

A British Standards booklet describes tests which can be performed in testing laboratories with purpose-made machines to compare how woods behave when—compressed; placed under cleavage pressure; indented with a steel ball; impacted by a dropping weight; subjected to shear forces; placed under bending forces; stretched; glue tests. Some of these are shown in Drawings 1 to 8.

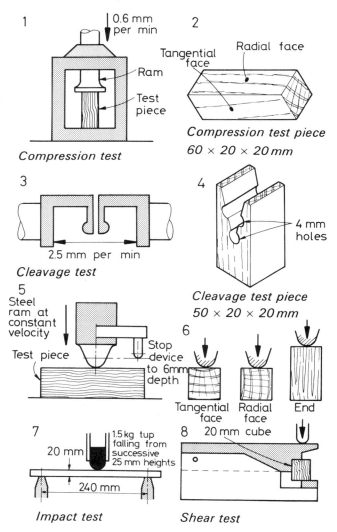

1 Compression test

2 Compression test piece $60 \times 20 \times 20\,mm$

3 Cleavage test

4 Cleavage test piece $50 \times 20 \times 20\,mm$

5

6 Tangential face / Radial face / End

7 Impact test

8 Shear test

Materials testing in test laboratories

The photographs on this page show some of the general materials testing machines in use under laboratory conditions. The five machines shown are only a small selection from those available.

The photographs were taken in the materials testing laboratory at the North Staffordshire Polytechnic with the kind permission of Mr. R. Preston, a lecturer at the Polytechnic, and the assistance of Mr. D. Cheers, a technician for the laboratory.

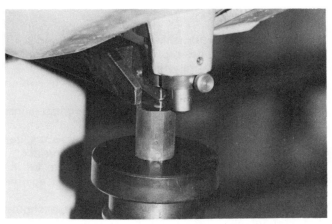

A Vickers hardness testing machine. The apex of a square pyramid made from a diamond is forced under known pressures into the material under test. The distances across the diagonals of the indentation are then measured.

A sample of steel under test for tension. The test sample is held in the vice jaws of the machine. This machine can also be used for compression testing.

A machine for measuring torsion. The material under test is being subjected to known torques.

The stress pattern for a shaped part displayed on polarised screens.

A sample of steel undergoing a fatigue test. Weights hung from a ball race at the end of the test piece subject the specimen to constantly changing stresses as the specimen rotates.

Surface finish

When considering the surface finishes required for a technology project design, you will probably only be considering those parts which:
1. can be seen—e.g. boxes or containers in which working parts are fitted;
2. require protection—e.g. against weathering; against penetration by water;
3. require special treatment—e.g. when you are deliberately attempting to produce a non-slip surface;
4. give rise to special factors—e.g. require to be insulated against heat; require to be insulated against electricity.

Many parts of technology designs may not, or should not, require any surface finish. In some cases, the application of a surface finish may be unsafe or may cause a component function to be affected. Such parts as the following do not normally require a surface finish:
1. Electrical and electronics components. Best leave well alone.
2. Mechanical working parts—may require lubricating from time to time, or some other form of maintenance such as adjusting, but rarely require a surface finish.
3. Some structural parts e.g. Meccano parts may be considered to already have a satisfactory surface finish.

Many of the materials used in technology projects may not need a surface finish. These include:
Plastics
Most plastics do not require an applied finish, although it may be necessary to paint some GRP mouldings (glass reinforced plastic).
Woods
Plastic faced laminates, such as 'Formica' and other melamine coated sheets which are glued or bonded to plywoods and chipboards.
Metals
Aluminium; stainless steels; zinc plated fittings and sheet material; cadmium plated nuts and bolts.

When, however, it is necessary to apply a surface finish to parts of a design for a technology project, the following may be considered. When using proprietary finishes, always read and then follow, the maker's instructions as to how the finish should be applied. Failure to read such instructions may lead to poor results.

Woods
Wax polish; french polish; polyurethane varnish; cellulose lacquer; teak oil; oil paints; polyurethane paints; emulsion paints; aluminium paint; stains; preservatives.
Metals
Oil paints; polyurethane paints; phosphating paints (anti-rust); aluminium paint; cellulose paints (can be applied as sprays); polyurethane varnish; plastic dip coating; aluminium anodising.

Fittings

Technology projects can involve electrical circuits; electronics circuits; structures; mechanisms; power supplies; pneumatics circuits; hydraulics; fluidics; aeronautics; research into sources of energy; materials testing; model making; among many other topics and systems requiring investigation. Each of the systems involved in this large number and variety of groups, will have had its own sets and types of equipment and fittings developed. In addition, other fittings can be involved in technology project work such as—nails; screws; nuts and bolts; rivets; hinges; pivoting devices; various clips; mounting boards. These involve another long list of fittings.

Because of this complexity of so many different types of equipment and fittings required for school technology, no attempt has been made to list them here. Reference books, catalogues and brochures can be purchased, or obtained from the firms who supply the fittings. Catalogues, in particular, can be a valuable source of reference, not only listing the fittings required, but also giving dimensions and other information. Some firms send out catalogues which also give details of how their fittings can be used to best advantage and also list such essential information as the limits of tolerance to power supply and so on.

Fittings lists
As part of an investigation you may well consider including lists of the fittings and components required for constructing technology design. These lists should then be included with the notes and graphics in the folder or folio in which details of an investigation are stored. Lists, such as that below, serve several purposes.

No.	Component	Description	Catalogue No.
2	Resistors	$10\,k\Omega$	142–485
1	Resistor	$2.7\,k\Omega$	142–407
1	Resistor	$2.2\,k\Omega$	142–390
1	Variable resistor	$50\,k\Omega$	161–717
1	LDR	ORP12	305–620
1	Filament bulb	6V 0.06A	586–172
1	Transistor pnp	2N 3703	294–334
1	Battery	9V PP9	591–095
1	Integrated circuit	741 Op Amp	305–311
1	Stripboard (copper strip)	$95\,mm \times 60\,mm$	from store

1. They help to ensure that essential items have not been forgotten or missed.
2. They can be used as reference material if one wishes to incorporate part of a design in another project.
3. They can be used as order lists for giving in at a school of college store or for sending to a firm when ordering supplies for project work.

A components list for a circuit to be incorporated in a project design is given on page 124. This is a detailed list of the components needed for a photo-switch circuit in which an integrated circuit operational amplifier (a 741 Op Amp) is incorporated.

Safety

When designing for school technology projects, the investigation frequently involves experimental work. The dangers involved in this part of the design process must always be borne in mind. Here we are concerned with the dangers arising within the actual design resulting from the investigation. The following are notes concerning some of the hazards involved. Some of the notes are in question form.

Fire
Are the materials involved in the design combustible? Should non-combustible materials be used throughout? Does the use of any combustible materials give rise to a fire hazard? Is the power of the project derived from a primary heat source e.g. town gas? If so, would a non-combustible source be preferable? If there is a heat output from the power source, is there adequate heat insulation?

Electricity
Are all parts fully insulated against leakage of current? If the electric power is from a mains source, is the design fully earthed? Is it necessary to use mains power? Can the project instead be adequately powered from a low voltage battery or via a transformer? Are capacitors in the circuit connected in such a manner that they do not discharge after a circuit is switched OFF? Is there a danger of water seeping into the circuit and so causing 'shorts'? Is the circuit fully safe? Is the circuit really OFF when switched off? When experimenting with any circuits always disconnect from the power supply if the experiment is left for any period of time, no matter how short.

Control systems
Do the control systems function as expected? Check and then re-check. There is always a danger inherent in any control system that is not designed so as to be 'fool' proof. A system should act in exactly the manner expected.

Structures
Are structures quite safe? Will they take the loads imposed on them? Did the investigation into the design of a structure examine all aspects of possible failure? Is a safety margin built into the calculations on which a structure depends? In industrial design a safety factor of many times the expected loading is always applied to structural design. The unexpected stress caused by suddenly increased weight or movement is a feature which should be examined when designing a structure. If a structure is of inadequate strength, even in model making, its failure can lead to accidents. An example   if a Meccano structure fails while carrying a motor which is running at the moment of failure, an accident may result.

Mechanisms
Does a mechanism work as expected? Test all movements, in all directions, before allowing a mechanism to run under power. Is there sufficient space for the mechanism to work? Is any part of it liable to fail? Is there a possibility for example that a slipping belt or a pulley or a gear running on insufficient power may cause a safety hazard? Are all parts of a mechanism properly secured to the structure supporting it? Can each part of a mechanism be properly serviced?

Speed
When designing for projects involving vehicles— whether on land, on water or in the air—there is always an inherent danger due to the speed at which a vehicle can travel, or the danger that their control systems are inadequate. Before allowing any such vehicles to operate under their own systems, they should always be tested on a tether or on a safety line without people being in the vehicle. In fact the testing by tether should be regarded as part of the design investigation. Testing of vehicles on water should always be carried out with great care if people are to be eventually carried in the vehicles. Life jackets must be worn by such people and others should be standing by to give any necessary assistance even after tests prove to be satisfactory.

Instructions
Always make quite certain that instructions for operating a design are accurate and clear. This applies to spoken or written instructions given to others who may wish to operate your designs. It also applies to instructions on controls and/or switch systems. If you show symbols on controls, be quite certain they convey exactly what the action of the controls should be.

Deterioration
Will the materials used in your design deteriorate? If so should they be changed? Have you allowed for maintenance of the design if parts do deteriorate? If deterioration may be caused by lack of lubrication of moving parts, has allowance been made in the design to ensure that lubrication can take place? Are those parts of a design which may be liable to become wet or dry, hot or cold, going to deteriorate as a result? Can you design so that such deterioration does not occur?

Fumes
Will the design give off fumes which may be dangerous? If so, have you made proper arrangements for the fumes to be circulated away from the people using it? Could the fumes be avoided by using different materials, different systems or a different power source?

Dangerous materials
Such materials as lead, asbestos, some oils, and others,

should be avoided because they are potentially dangerous to human beings under some circumstances. Some plastics foams—polyurethane foam such as is used for some upholstery—can be a fire and fumes risk if there is the slightest danger of their overheating and catching fire. Some sprays are best avoided. Lead-free paints should be used where possible. Some resins can give off fumes when heated.

Pneumatics

Compressed air should be regarded as an extremely dangerous element in school technology. Check *all* pipelines to ensure they are secure. Ensure that no loose pipes are connected to the supply. A loose open-ended pipe can act like a dangerous whip when compressed air is switched into it. Check air pressures and never use your fingers to check the air pressure in a cylinder—always check with gauges. Small compressed air cylinders should be regarded as dangerous and used accordingly.

Economics

The economics of designing for school technology can be considered under four headings:
1. The costs in money, involved in purchasing the materials, fittings and components involved in the design.
2. The amount of time spent on designing and making the completed design.
3. The economical use of space.
4. The economical use of the power which may be consumed in order that the design functions well.

Money costs

Projects undertaken in school technology can vary a great deal in their actual money costs. Some can be regarded as relatively cheap. Some may involve practically no cash outlay at all, as for example when components are stripped from secondhand equipment or when fittings are re-used after having been involved

in other project work. Many projects can, however, be regarded as being expensive. Whether cheap or expensive, each design should be costed. One way of costing is to write a list of all the materials, components and fittings required and find what each costs from relevant catalogues. One such list for a project which includes a simple pneumatics circuit is given below.

Note—The costs shown are based on those quoted in catalogues at the time of writing. The reader may well find these costs to be different at the time of reading.

Time

In schools and colleges the time spent on designing and making a technology design can be regarded as being extremely well spent in an educational sense. A good plan, however, is to make a note of the time spent on designing and making a project and making notes of the total time spent, to be kept among other notes in a design folder.

Space

Economy in the use of space should be regarded as an essential feature to be investigated when designing. From the point of view of economy of materials and hence of costs in money terms, there is no need to make any design larger than necessary. When the microscopic sizes of modern circuits involved in 'silicon chip' designs are considered, the importance of economy of space in designing can be better understood.

Power

With the increasing need to conserve the world's energy resources as represented by our fossil fuels, the need to reduce power consumption to a minimum takes on a new interest. Although the actual amounts of power consumed by any of the school technology projects will be relatively small, a good design aim is to design so as to consume as little power as possible to gain a maximum effect.

In connection with the conservation of the world's energy resources some interesting school technology projects can be devised based on obtaining usable energy from the sun, from the wind, from rising air and from other 'natural' resources.

No.	Item	Dimensions	Cost
2	3-port valves		4.50
1	5-port valve		4.80
1	Flow restrictor		3.60
15 metres	Plastic piping	3 mm diameter	2.25
2	Tee joints		1.90
1	Mild steel strip	120 mm × 40 mm × 3 mm	
1	Mild steel strip	60 mm × 20 mm × 3 mm	1.25
1	Mild steel rod	50 mm × 20 mm diameter	
1	Mild steel plate	120 mm × 30 mm × 16 SWG	
1	Hardwood	600 mm × 40 mm × 8 mm	75
1	Plywood	150 mm × 100 mm × 3 mm	
Various	Screws, pins, nuts and bolts		50
		Total cost	£19.55

Bibliography

Some of the books given in this bibliography are text books written for school pupils and college students. Some have been written for teachers and lecturers, some are reference books, others are general text books. No attempt has been made to classify these books into specific reader groups.

Design and Craft
The attention of users of this book is drawn to another book by the same publisher:
Design and Craft by A. Yarwood and S. Dunn.

Design and Technology
Design and Technology in the School Curriculum, Tom Dodd. (Hodder and Stoughton).
School Technology in Action, ed. Alan R. Marshall. (Hodder and Stoughton).
Presentation of Design, T. Pettit. (Edward Arnold).
Design Education at Secondary Level—A Design Council report. (The Design Council).

Schools Council Project Technology
Published by Hodder and Stoughton:
Control Technology, Pupil's Assignments.
Control Technology, Follow-up sheets.
Control Technology, Teacher's Book.
Photocell Applications.
Basic Electronics
> Book 1   Introducing electronics. Measuring instruments.
> Book 2   Resistors in circuits. Capacitors in circuits.
> Book 3   Inductors in circuits. Diodes in circuits.
> Book 4   Meters. Voltage dividers.
> Book 5   Transistors in circuits. Transistors in action. Post-transistor projects.

Basic Electronics. Complete volume of all 5 books.

Published by Heinemann Educational Books:
Simple Bridge Structures.
Introducing Fluidics.
Gas-Fired Muffle Furnaces.
The Ship and Her Environment.
Design with Plastics.
Industrial Archaeology for Schools.
Food Science and Technology.
Basic Electrical and Electronic Construction Methods.
Simple Computer and Control Logic.

Published by Oliver and Boyd:
Modular courses in technology – each title published as a Teacher's Guide and as a Workbook.
Energy Resources.
Electronics.
Mechanisms.
Structures.
Problem Solving.
Materials Technology.
Pneumatics.
Instrumentation.
Technology and Society.
Electrical Applications.

British Standards publications
The British Standards Institution (BSI) publishes the *British Standards Year Book* annually. This lists all publications of the BSI with a short description of the contents of each.

British Standards publications which may be of value to those practicing school technology are listed below. BSI publishes some 9000 Standards together with a number of Published Documents (PDs) and other publications. Many of these can provide valuable and accurate background material for school technology.

BS: 308	Engineering drawing practice.
BS: 499	Specification for symbols for welding.
BS: 1192	Building drawing practice.
BS: 2917	Specification for graphical symbols used on diagrams for fluid power systems and components.
BS: 3643	ISOmetric screw threads.
BS: 3939	Graphical symbols for electrical power, telecommunications and electronics diagrams.
BS: 4058	Specification for processing flow chart symbols, rules and conventions.
BS: 4163	Recommendations for health and safety in workshops of schools and colleges.
PD: 7300	Nuts and bolts: recommended drawing ratios for schools and colleges.
PD: 7301	Graphical communication: a teacher's pack.
PD: 7308	Engineering drawing practice for schools and colleges.

Complete sets of all the British Standards Institution's publications are held at a number of libraries and public buildings throughout the British Isles and in other libraries and public buildings throughout the world. A list of these complete sets, which are made available to the public for inspection and study, can be found in the current British Standards Year Book.

Graphics
Geometrical and Technical Drawing Books 1 and 2, A. Yarwood. (Thomas Nelson).
Graphical Communication Books 1 and 2, A. Yarwood. (Thomas Nelson).
Design Drawing Books 1, 2 and 3, John Rolfe. (Hodder and Stoughton).

Computers
Microfuture by John Shelley. (Pitman).
Computer Studies—A Practical Approach by G. M. Croft. (Hodder and Stoughton).
Comprehensive Computer Studies by P. Bishop. (Edward Arnold).
Microelectronics and Microcomputers by L. R. Carter and E. Huzan. (Teach Yourself Books/Hodder and Stoughton).

Computer Programming in BASIC by L. R. Carter and E. Huzan (Teach Yourself Books/Hodder and Stoughton).
Computer Programming in COBOL by M. Fisher. (Teach Yourself Books/Hodder and Stoughton).
Computer Programming in FORTRAN by A. S. Radford. (Teach Yourself Books/Hodder and Stoughton).
Computer Programming in Pascal by D. Lightfoot. (Teach Yourself Books/Hodder and Stoughton).
How it Works—The Computer. (Ladybird Books).

General
Ordinary Level Physics by A. F. Abbott. (Heinemann).
Theoretical Mechanics by G. S. Light and J. B. Higham. (Longman).
Applied Mechanics by J. D. Walker. (Hodder and Stoughton).
Structural Mechanics by Trefor J. Reynolds and Lewis E. Kent. (Hodder and Stoughton).
The *How it Works* series published by Ladybird Books.
How it Works—The Motor Car
How it Works—The Rocket
How it Works—The Aeroplane
How it Works—Television
How it Works—The Locomotive
How it Works—The Hovercraft
How it Works—The Camera
How it Works—Farm Machinery
How it Works—The Telescope and Microscope
How it Works—Printing Processes
How it Works—The Telephone
Electronics by W. P. Jolly. (Teach Yourself Books/Hodder and Stougton).
The Challenge of the Chip. (HMSO).
Adventures with Electronics by Tom Duncan. (John Murray).
Adventures with Microelectronics by Tom Duncan. (John Murray).
Adventures with Digital Electronics by Tom Duncan. (John Murray).
Snaps and Circuits by Frettsome and Morris. (Hart Davis and Hutchinson).
Elementary Electronics by Mel Sladdin. (Hodder and Stoughton).
Pedal Power by James C. McCullagh. (Rodale Press) (USA).
110 IC Timer Projects by J. H. Gilder. (Newnes Technical Books).
Operational Amplifiers: Their Principles and Applications by J. B. Dance. (Newnes Technical Books).
Electronics Options by P. Gormley, J. S. Hagan and C. C. Atkinson. (Edward Arnold).
Radio Spares Components Catalogue. (RS Components Limited). (A valuable reference book).

The Dewey system of book subjects classification
Non-fiction books in libraries are usually classified by subjects in a system known as the Dewey system, sometimes called the Dewey decimal system. As a guide only, the following list may assist the reader when seeking books from which to gain information about details when designing for a technology project. The complete Dewey system is so vast and complex that the following list can only be taken as a guide. Further information concerning the Dewey classification can be found in the cataloguing system in public, school and college libraries.

Subject		Dewey subject number
Aeroplanes	General	629.13
	Aeromodelling	629.1
	Aeronautics	629.1
Bridges		624
Cameras		771.3
Chemistry		540
Computing		001.64
		510.78
		510.81
Design (Craft)		745.4
Electric circuits		537.61
		621.31
Electronic circuits microelectronics		621.381
Energy conservation		333.72
Fluidics		629
Graphic design		741.6
Hydraulics		621.8
		629.8
Hovercraft		629.3
Materials Engineering		620.1
	Conservation	069.4
Mathematical logic		511.3
Mechanics		620.1
Microscopes		535.332
Plastics Technology		668.4
	Education	745.57
Rockets		623.45
Structures Engineering		624.1
Technical Drawing		604.2
Technology Careers		602.3
	Research	607.2
Telephones		621.385
Telescopes		621.38
Television		621.388

Solutions

Probably the best method of showing, and making records of, the development of ideas and suggestions for solving the design problems associated with technological projects, is by the use of drawings associated with the necessary written notes. This has already been clearly illustrated in pages 8 to 44 in which ten examples of school technology projects have been given. Solutions drawings may be made freehand or with the aid of instruments, in pencil or in ink. Colour may or may not be included.

Drawing papers

Practically any paper is suitable for the recording of drawings for solutions of technology projects, although the use of good quality drawing paper is advisable. Many different sizes of paper are available but the 'A' series of sheet sizes are now in common use in schools. A size sheets, particularly A3, will be found to be suitable when drawing for technology. Two types of grid paper, shown in drawings on this page, can also be of value as aids to good drawing. These two papers are isometric grid and square grid papers. The grid lines are printed in green or in blue with either 5 mm or 10 mm grid spacing. Thus in the case of isometric grid papers, each triangle side is 5 or 10 mm long. In the square grid papers, the squares are of 5 or 10 mm edges. Other grid spacings are also available.

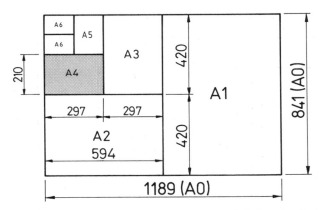

Sizes of the A series of drawing sheets. A4 size shaded. Dimensions in millimetres.

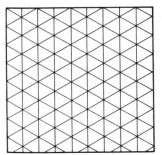

5 mm isometric grid sheet

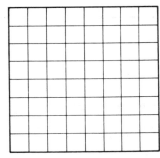

5 mm square grid sheet

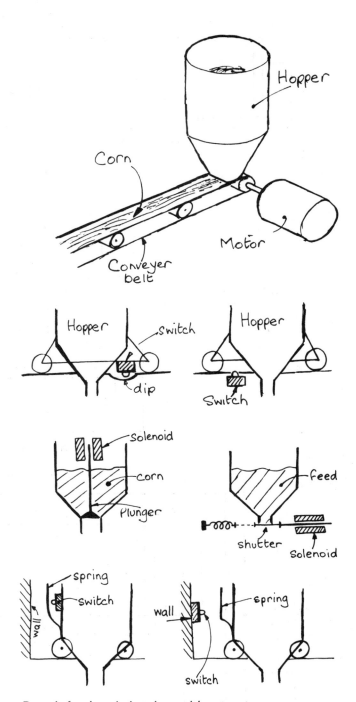

Rough freehand sketches without notes

Methods of drawing

A number of methods of drawing are shown here. These include orthographic projection, oblique cabinet projection, isometric drawing, planometric drawing and perspective drawing. These methods may be employed in instrument-aided or freehand drawing. The student should choose the methods best suited to the graphical description of the problems involved in the technology project being undertaken.

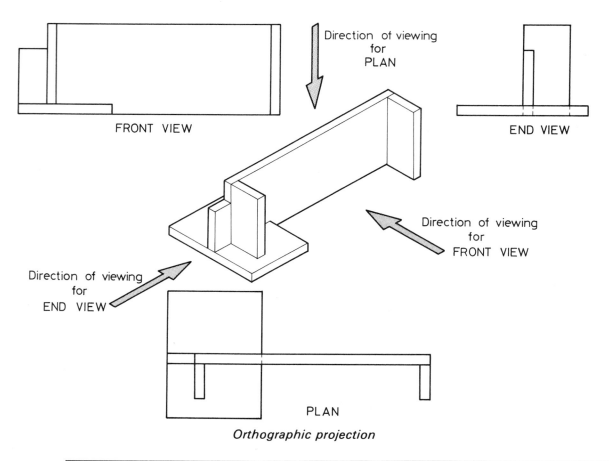

FRONT VIEW

Direction of viewing
for
PLAN

END VIEW

Direction of viewing
for
FRONT VIEW

Direction of viewing
for
END VIEW

PLAN

Orthographic projection

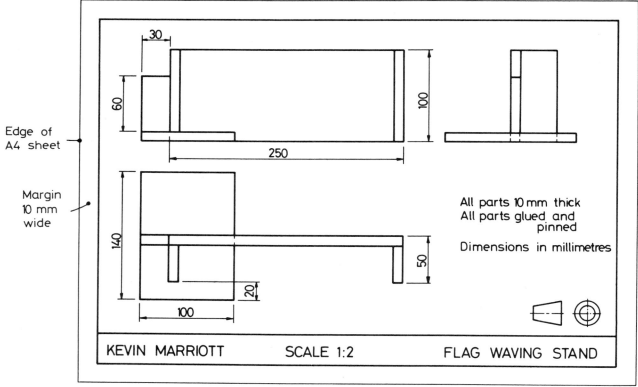

Edge of
A4 sheet

Margin
10 mm
wide

30

60

250

100

140

100

20

50

All parts 10 mm thick
All parts glued and
pinned

Dimensions in millimetres

KEVIN MARRIOTT   SCALE 1:2   FLAG WAVING STAND

The three views laid out on an A4 sheet

Orthographic projection

The drawings on page 130 opposite show the wooden stand which forms part of the project already described on pages 30 to 33.

Upper drawing

The object to be drawn is looked at from its front and what is seen is then drawn as a *front view*. The object is then viewed from an end and what is seen is drawn as an *end view*. A viewing position is then taken from above and what is seen is drawn as a *plan*.

Lower drawing

The end view and front view are placed in line with each other and the plan placed immediately below the front view. Such a drawing is said to be a first angle orthographic projection. Another type of orthographic projection, known as third angle, is not dealt with here. In the lower drawing note the following:

1. The drawing has been made on an A4 size sheet of paper.
2. A margin, 10 mm from the sheet edges.
3. A 'title block', which includes the student's name and the title of the drawing, together with the scale to which the drawing has been made.
4. The drawing is fully dimensioned. Sufficient dimensions should be included to allow the object to be made by 'reading' the drawing.
5. In the bottom right hand corner of the drawing, a symbol denoting First Angle projection has been drawn.

How many views?

When employing orthographic projection methods of drawing, the decision as to how many views to include must be considered. The determining factor in making this decision is—what is the least number of views which will fully describe the object being drawn?

In the case of the flag waving stand shown on page 130, three views are needed to fully describe the shapes of its various parts. Only two views—front and end views—are needed to describe the air conditioning unit box shown on this page. Single view drawings—front view only—can often be used to describe flat articles or to show how parts function in a technological project. Two examples of single view drawings are shown.

Square grid papers are very suitable for orthographic projection drawings. The layout of switches on a panel shown here has been drawn on a 10 mm square grid. The grid assists in accurate layout and the sizes and positions of the various parts of the drawing can be found by counting the squares, rather than by placing dimensions on the drawing. Orthographic projection drawings on plain paper should be dimensioned, if the object being drawn is to be made from the drawing.

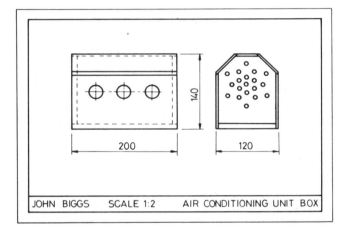

An example of a two-view orthographic projection

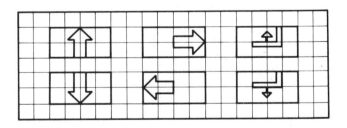

An example of a front view drawn on a square grid—layout of switches for a fork lift truck project

Single view drawing of a coupling for a window cleaning unit

Isometric drawing

Isometric drawing is a method by which a pictorial view of an article can be produced. A ruler or a T square and a 30°, 60° set square are required. The method of constructing an isometric drawing is simple and easy to understand. Providing the article being drawn is not particularly complex in shape, a good pictorial view can be easily drawn using this method.

Stages in producing an isometric drawing

1. Draw a line at 30° to the horizontal with the aid of a 30°, 60° set square. Measure the length of the item to be drawn along this line—either the full length of a full size drawing is being constructed or the scaled length, if a scaled drawing is to be made.

2. Draw a vertical line with the aid of the set square. Measure the vertical height of the item along this line.
3. Draw a second 30° line and measure along it the depth of the item.
4. Complete the required isometric drawing with the aid of the set square.

Isometric drawing on isometric grids

An example of an isometric drawing on a 10 mm isometric grid is shown. Such a drawing can be constructed on the grid paper without the aid of a set square. The drawing can then be cut out and pasted up alongside other drawings in a project folder.

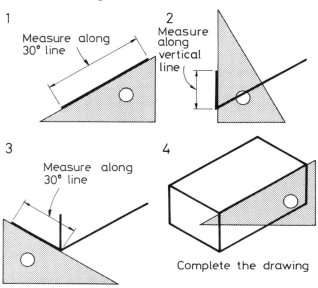

Stages in producing an isometric drawing

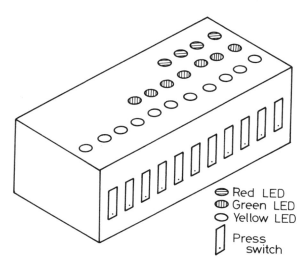

Isometric drawing of control box for an electronics combination lock

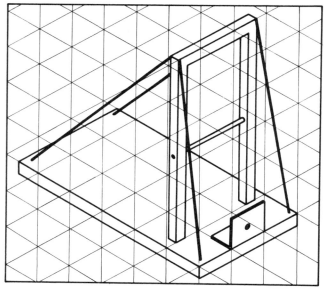

Isometric drawing of an experimental automatic barrier gantry. Drawn on 10 mm isometric grid paper

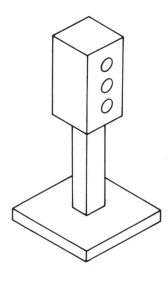

Compare this isometric drawing with the oblique cabinet projection of the same model traffic control system shown on page 133

Oblique cabinet drawing

Another way of producing pictorial views of items is by the use of the oblique cabinet drawing method. This method is particularly suited to the production of drawings for technology projects because of its simplicity.

Stages in producing an oblique cabinet drawing

1. Draw a front view of the article.
2. Draw lines from the front view at 45° with the aid of a 45° set square. Along these 45° lines take measurements at half the scale used for the front view. Thus if the front view is drawn full size, half size dimensions are taken along the 45° lines.
3. Continue taking 45° lines from all parts of the front view.
4. Complete the drawing with the aid of the set square.

Note

The half scale measurements taken along the 45° lines are necessary in order to avoid the distortions of depth which occur when full size dimensions are taken along the sloping lines.

Planometric drawing

This is yet another method of drawing with the aid of set squares. Planometric drawing is used mainly to show room and building layouts. An example of such a room layout connected with a technology project is shown.

Two methods, one involving the aid of a 45° set square, the second involving the aid of a 30°, 60° set square, are in common use. In the case of the 45° set square method, normal practice is to take the same scale measurements along all verticals and both 45° axes. When using the 30°, 60° method common practice is to draw verticals at about $\frac{3}{4}$ or $\frac{2}{3}$ the scale used on the 30° and 60° axes.

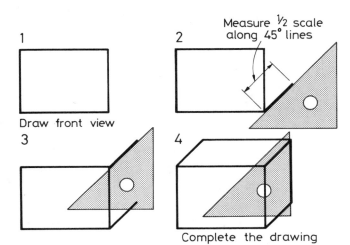

Stages in producing an oblique cabinet drawing

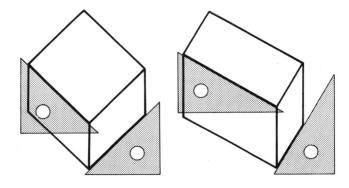

Two methods of planometric drawing

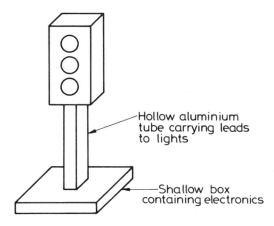

An example of oblique cabinet drawing. Model traffic control system

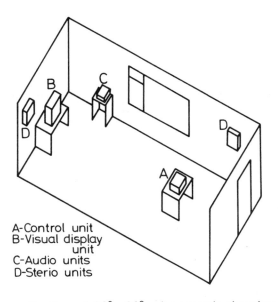

A-Control unit
B-Visual display unit
C-Audio units
D-Sterio units

An application of 30°, 60° planometric drawing

133

Estimated perspective drawing

Some of the drawings associated with the solutions to a technology project may be best provided by using perspective drawings. Many methods of producing perspective drawings can be found in text books but, for the purposes envisaged here, the best methods are those of one-point and two-point estimated perspective.

One-point estimated perspective

Draw the front view of the item being portrayed. Estimate the position of a single vanishing point (VP) either above the front view or to either side. Draw lines from the front view to the vanishing point. Take measurements along one of these sloping lines, remembering to reduce the scale to avoid a too-deep appearance of the pictorial view. All verticals are drawn upright.

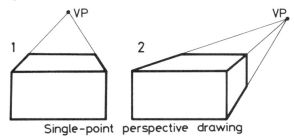

Single-point perspective drawing

Methods of estimated perspective drawing

Two-point estimated perspective

In this method estimate the positions of two vanishing points. They should be horizontally in line with each other. Draw the vertical line closest to the viewer and from this take lines to the vanishing points. Remember to reduce the scale along the sloping lines to avoid depth distortion.

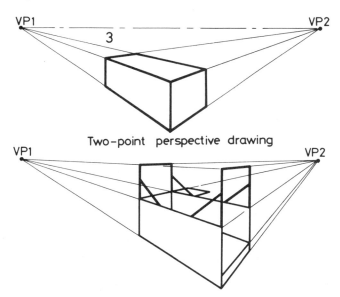

Two-point perspective drawing of a gantry and water tank for a model lock gate project

Sectional views

When drawing sectional views, an imaginary cut is made through the article and the cut surface then drawn. This drawing technique is of particular value when making drawings for technology projects because the shape and construction of the interiors of objects can be clearly shown in sectional views.

Drawings 1, 2 and 3 show the method. Drawings 1 and 2 are isometric drawings and Drawing 3 is a two-view orthographic projection, one view being a sectional view. Two examples of the use of sectional views in technology projects are also given.

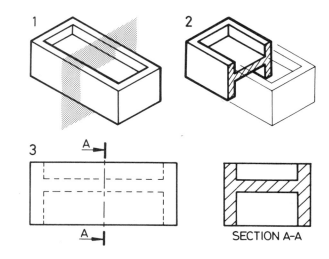

Sectional views in a two-view orthographic projection

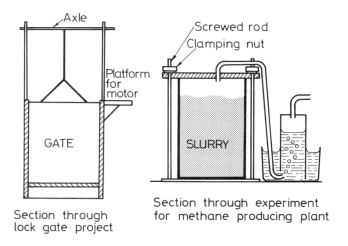

Section through lock gate project

Section through experiment for methane producing plant

Examples of sectional drawings

Developments

In some technology projects, parts of a completed design may be made from sheet materials. In such cases a *development* of the surfaces of the material required for the part may be necessary. Two examples of developments, both for the bubble tank shown in Drawing 1, are shown. Drawing 2 is a development for the main part of the tank. This drawing has been made with drawing instruments. Drawing 3 shows the development of the pipe support of the bubble tank drawn on 5 mm grid paper.

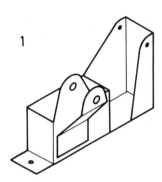

Isometric drawing of bubble tank

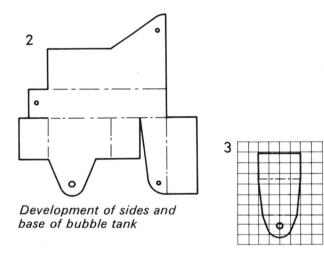

Development of sides and base of bubble tank

Development, drawn on 5mm square grid, of pipe support of bubble tank

Freehand drawing

Freehand drawing is of particular importance when producing the graphics required for solving technology designs. Any of the instrument aided methods of drawing can be applied when working freehand. Five examples are given on this page. These are:
Freehand three-view orthographic projection
Freehand isometric drawing
Freehand oblique cabinet projection
Freehand planometric projection
Freehand two-point perspective drawing.

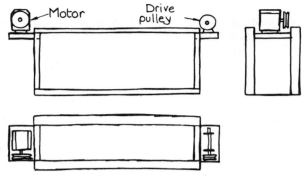

Farm effluent collection trench

An example of freehand orthographic projection

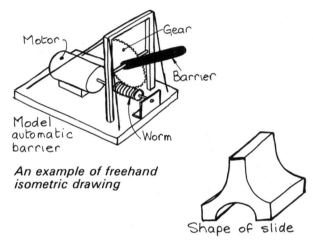

An example of freehand isometric drawing

An example of freehand oblique cabinet projection

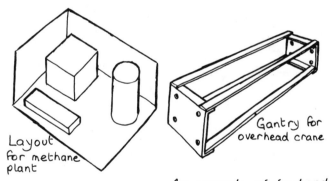

An example of freehand planometric projection

An example of freehand two-point perspective drawing

Scribble drawings

Many of the solutions to technology project designs may take the form of drawings scribbled 'on the back of an envelope'. It is as well to keep such drawings on record no matter how crudely drawn they may be. They can be pasted into or 'Sellotaped' into folios and folders containing the investigation for a technology project.

135

Exercises

1. A room in a museum contains many priceless exhibits which must be protected against intruders.
 (i) Select one method from the list below and explain how it could be incorporated in a suitable warning device.
(a) Heat change
(b) Mechanically operated switches
(c) Air movement
(d) Light sensors
(ii) The system must be 'failsafe', i.e. when a fault in the equipment occurs, an alarm is given in such a manner that it is not confused with the intruder alarm.
Explain how a 'failsafe' mechanism could be incorporated in your device.

2. Sketch and label the following basic mechanisms:
(a) a ratchet;
(b) a cam and follower;
(c) a rack and pinion;
(d) a friction clutch   (SUJB)

3. (a) The diagrams in Drawing 1 show three different applications of levers. Which diagram shows:
 (i) a class one lever,
 (ii) a class two lever,
 (iii) a class three lever?
(b) Drawing 2 shows a linkage that will reverse the direction of its input motion.
Sketch a linkage that produces an output motion in the same direction as the input motion.   (SUJB)

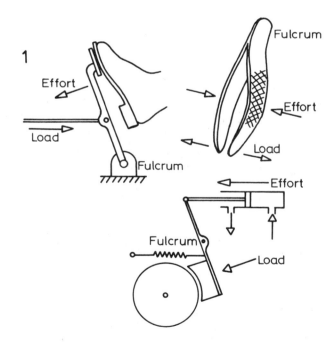

(a) brake pedal
(b) a pair of tweezers
(c) a pneumatic brake

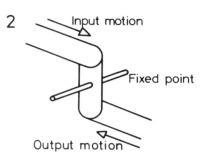

4. Drawing 3 shows the side of a storm water culvert with a pressure retaining plate of 600 000 sq mm which will retain an overall water pressure of 6750 Nm^{-2}. Should the water surface rise above the level shown the gate will open.
(a) Design a lever system that will apply a force equal to the maximum water pressure and allows the gate to open when the pressure rises.
(b) Show by calculation the amount of loading your system will require.

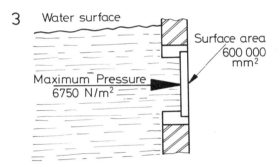

5. The aerofoil shown in Drawing 4 was placed in an air flow. The air pressures were compared at various points on the aerofoil's surface using a multiple manometer similar to that shown in Drawing 5. Tube 1 was connected to position 1 on the aerofoil, tube 2 to position 2 etc.

(a) Copy the diagram of the manometer on to your answer sheet and mark on it the levels of the fluid when the aerofoil was at an angle of attack of about 12°.
(b) The angle of attack was then slowly increased until a state of stall was reached. What indications would there be from the manometer levels that the aerofoil was:
 (i) entering a state of stall,
 (ii) in a state of stall?   (SUJB)

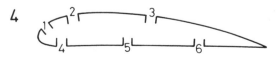

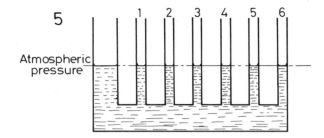

5

Atmospheric pressure

the vertical ties between each of the frames and the ceiling at A and B respectively.

Calculate by graphical methods—

(i) The forces in each of the five rods.

(ii) Which rods are in tension and which in compression.

(iii) The forces in the vertical ties between joints and frames at C and at D.

6. In the study of the mechanical properties of timber, a testing device is required by which the stiffness and elasticity can be tested on samples of timber of dimensions 500 mm by 50 mm by 25 mm. Loading to the centre of the specimen is to be by means of a system of levers—Drawing 6.

Design such a testing rig that will apply a force to the specimen by means of a system of levers and has a method of measuring the amount of deflection.

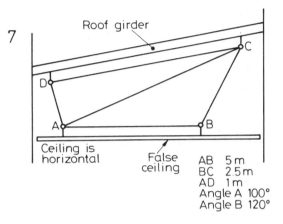

7

Roof girder

Ceiling is horizontal   False ceiling

AB 5 m
BC 2.5 m
AD 1 m
Angle A 100°
Angle B 120°

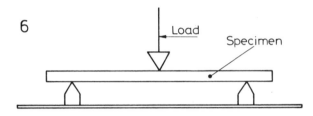

6

Load   Specimen

7. Design an apparatus for determining the speed with which a pellet leaves the barrel of an air gun. For the purposes of the design, assume that the pellet has a mass of 5 g and that a typical speed is 400 m/s. Explain how the speed is calculated from the measurements made with the apparatus.   (Oxford)

8. Describe and explain a possible design for the control circuit of a pedestrian crossing which has the following specification:

(a) When a push button switch is operated the traffic light goes to red and the pedestrian light goes to green.

(b) After ten seconds the traffic light goes to green and the pedestrian light goes to red.

(c) The system is now ready to be used again.

A real life system built to the above specification would have a number of unsatisfactory features. Rewrite the specification to overcome these unsatisfactory features, and give circuit diagrams and notes for a possible design to satisfy this new specification. Describe how your proposed circuit operates. (AEB)

9. A false ceiling is suspended from the girders of a roof by a series of pin-jointed frames made from light rods as shown in Drawing 7. In order to maintain the ceiling in place forces of 1500 and 2000 newtons are required in

10. As part of an investigation of the effects of insecticides, an agrochemical company has decided to study the harvest mice in a cornfield.

Design a trap to catch a harvest mouse without harming it so that the harvest mouse may be released after observations have been made on it. Assume that a suitable bait is available and that a harvest mouse has a body of length 60 mm and a tail of length 80 mm.

(Oxford)

11. (a) A small machine is needed to clean the floor of a highly radio-active research area which is rectangular in shape. Describe a possible design for the machine and its control system, so that the operator can control the machine from a separate room. The operator must be able to:—

i) Direct the machine to any part of the room i.e. he must be able to steer it and reverse its direction of movement.

ii) Switch on and off rotating brush heads and a vacuum cleaning device to collect the dust.

(b) It is required to redesign the machine so that it operates entirely automatically to allow the operator to set the machine in motion and leave it. Write a specification for the automatic version of the machine and give a possible design to satisfy this specification. Describe how your proposed system works.   (AEB)

12. Design a simple bench-mounted impact test device to measure the degree of hardness of case-hardened components. The impact load delivered should be constant and the indentation produced should be easily measurable.   (Oxford)

13. A small weather station is to be set up on the roof of a building. Measurements are required for wind speed and the daily amount of sunshine.
(a) Suggest methods by which the information may be obtained and transferred to a ground floor office. These readings are to be recorded in such a manner that comparisons can be made. Give details and drawings showing the equipment required and how each part works.
(b) Choose another climatic variable which can be measured and recorded. Give details of how this variable could be so measured and recorded.

14. As more engineering processes become automated it is necessary to design systems that automatically measure the size of machined components.

Design and sketch a device that would enable the operator of a grinding machine to measure the diameter of a workpiece whose final diameter should be 10.00 mm with a tolerance of ± 0.01 mm. The workpiece should not have to be removed from the grinding machine and so the measuring device should be attached to the machine. Clearly indicate how the device works. (SUJB)

15. (a) Drawing 8 represents a compound gear train.
(i) Find the gear ratio of the compound train.
(ii) If gear A is driven at 120 rev/min in a clockwise direction as indicated, what is the speed and direction of the shaft of gear D?
(b) (i) Sketch a mechanism that will allow a shaft to be rotated in one direction only.
(ii) A gear wheel is to be fixed to a shaft so that it can move along the shaft but not rotate about the shaft. Show how this can be done.
(c) A toy car is to be powered by an electric motor that drives the back axle.
(i) Sketch in outline a suitable gear system.
(ii) What information would be needed to calculate the speed of the car? Explain.
(d) Describe a quick-return mechanism, and state one application of it. (Oxford)

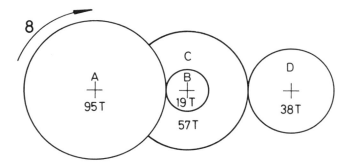

8

16. Drawing 9 shows instruction signs that are to be displayed to motorists from a remotely controlled display, positioned above each carriageway of a motorway.
(a) Describe *three* alternative ways in which these signs can be displayed giving *one* limitation in each case.

(b) Choose a best solution giving *five* reasons for your choice.
(c) State *three* ways by which you would evaluate your solution.

(SUJB)

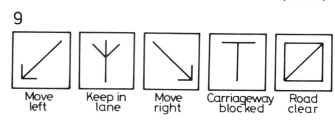

9

| Move left | Keep in lane | Move right | Carriageway blocked | Road clear |

17. Drawing 10 shows a cross-section through a fibre-optic cable.
(a) Which material, A or B, has the higher refractive index?
(b) What is the name of the phenomenon that fibre-optic cables use to trap the light inside the inner core?
(c) Draw a sketch showing how light is transmitted through a fibre-optic cable.
(d) Give *one* important use that is made of fibre-optics.

(SUJB)

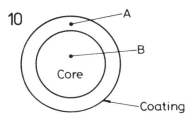

10

18. Drawing 11 shows a circuit often used for plotting transistor characteristics in common emitter connection. Study the circuit and answer the questions.
(i) What do the letters e, b and c stand for?
(ii) Which type of transistor is being investigated?
(iii) The two ammeters to be connected at X and Y are a microammeter and milliammeter. Which would you insert at X? Briefly explain why.
(iv) What is the purpose of R2?
(v) Why is it necessary to connect another resistor R1 in series with R2?
(vi) Give a reason why the voltmeter at Z should have a high resistance.
(vii) What does R3 control? What is the name given to a resistor used in this manner?
(viii) The transistor is supported in a clip or inserted in a cylindrical hole in a block of copper. Explain why it is desirable to do this.

(ix) Sketch the shape of the output characteristic that would be obtained with base current constant.

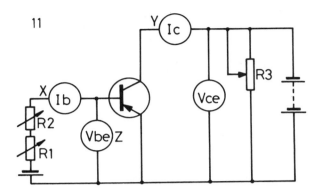

11

19. A pupil was asked to make a thermostat/flasher unit which would run on a 240 V mains supply. He constructed the piece of equipment shown in Drawing 12.
(a) Describe in detail how the unit would work.
(b) In order to increase the flash rate of the lamp it was suggested that the outside of the metal can is painted black. Explain why this idea might work.
(c) Suggest another way in which the flash rate might be increased.

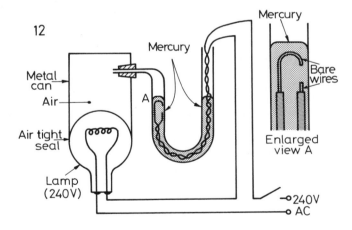

12

20. Drawing 13 shows the layout of a single track railway along which trains pass from A to B. At the start of the track is a light signal consisting of a red and a green light. Also along the track is a level crossing and a swing bridge over a river. Before a green light signal is given at A the following conditions must be met:
(a) The bridge is closed to river traffic and open to rail.
(b) The level crossing is closed to road traffic and open to rail.

(c) No other rail traffic is in the section A to B. You can assume that the rails are electrically insulated and that the rolling stock has metal wheels and axles.
(d) Should a fault occur in the signal system the red light is to remain on. The system is to be failsafe.
Design an electronics circuit which could be installed in a system so as to satisfy the conditions (a) to (d).

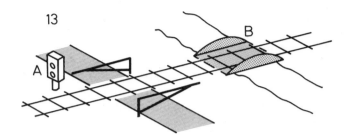

13

21. Drawing 14 shows a light and its switch inside a garage. When a car is driven into the garage during the hours of darkness, the driver must switch the light on and then, 30 seconds later, switch it off.
(a) Draw a wiring circuit for the lamp and its switch.
(b) Sketch and explain the outline of *two* systems which, when the switch is switched *on*, automatically switches *off* after a delay of 30 seconds. Any of the methods—electronics, pneumatics, a small motor or mechanical methods—can be employed.
(c) Select *one* of your two systems and explain how it functions in detail.
Note—Mains voltage is involved.

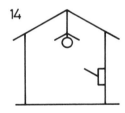

14

22. An enthusiast is planning to set up a small 'do it yourself' workshop for his own use. State *four* considerations that he should take into account before purchasing any equipment for the workshop.
(SUJB)

23. State *three* possible sources from which you could obtain information before you begin your major project.
(SUJB)

24. In the manufacture of computer components, it is necessary to count small circuit boards as they pass down a production line. It is important that the components are not subjected to electric or magnetic fields at this stage of their construction. Design a system that will both count the components as they pass down the production line and also sound a short audible alarm when 1000 components have passed by.
(Oxford)

Models

The construction of models should be regarded as an essential part of an analysis and investigation into the solving of a technology design brief. At some stage during an investigation, it will be necessary to test whether possible solutions are satisfactory or whether they function properly. The making of a model will often bring to the notice of the designer defects or faults in the manner in which the solution has been planned. Such defects can be then remedied as a result of further investigation after testing a model. Making and testing a model is quicker and cheaper than to make a completed design, only to find it does not function as expected. At some stage during an investigation unexpected problems can arise. If a model is made and then tested, solutions to the problems may be found as a result of the tests.

In school technology, the making of a design in scale model form can be the final aim of a design brief for a project. In many technology projects of this type, the making of a full size design may be beyond the capacity and expertise available in a school. However, even when a project is concerned only with the designing of scale models, other 'working' models may be needed on which to test solutions to problems which arise at the investigation stage of the design process.

Types of models for technology project design

A variety of different types of models may be made in connection with the design work for technology projects. Among others, the more common types are:

Electrical or electronic model circuits

These are usually constructed from kits to test whether a circuit which has been designed will operate as expected. Model circuits of this type are very common in designing for school technology. These kit constructed circuits are not necessarily small scale models, but are models in the sense that they are used for testing.

Working mechanical models

Working models can be constructed using kits such as Meccano, Fischertechnik or Hybridex, or may be constructed from wood, metal or plastics or any other materials at hand. Working models are made in order to test whether moving and working parts of a design such as link mechanisms, pulleys, gears and pistons function properly within the area for which they have been designed.

Models to check shape and form

Some models, with no working parts may be made to check whether their shape and form are suitable. They may also be made to check whether parts of a design fit each other, either physically or visually.

Mock-up models

Many models may be made from scrap materials or from very cheap materials, in order to judge final appearance of a suggested design or to test whether the model fits into an assembly or is suitable to fit with other parts within a layout.

Small scale models

The majority of models are made to a scale much smaller than full size, in order not to waste materials and because the smaller parts involved can be made more quickly than full size parts. Scale models economise in both time and materials. They are of particular value when testing working parts or when judging whether a design fits with other features in a layout.

Full size models

These are usually 'mock-ups' made to judge finished size and appearance.

Materials

A variety of different forms of 'kits' for making model electrical and electronics circuits as well as 'kits' for making mechanical models, have been shown throughout this book. In addition to these 'kits', a wide range of materials of all kinds will be found to be of value when making models for technology designs. Scrap materials or small pieces of material which would often be thrown away, may be found to be suitable for some model construction.

Materials suitable for frameworks of models

Wood strips glued and taped with masking tape; wood dowels. Thin metal rods; wire; soft iron wire; fuse wire. Electric cable. Plastic insulating piping. String. Thread. Meccano strip. Strips of plastic, e.g. acrylics, ABS. Strips of cardboard bent centrally to form a type of 'angle iron'.

Materials suitable for solid parts of models

Blocks of wood. Plastic bottles and other containers. Food containers both cardboard and tinplate. Rigid polyurethane foam. Expanded polystyrene foam. Clay. Plasticine. Plaster. Cement. Plastic rainwater piping. Plastic guttering.

Materials suitable for surfaces of models

Cardboards. Papers. Corrugated packing boards. Hardboard. Chipboard. Plywood. Sheet plastics. Aluminium foil. Thin mild steel sheet. Tinplate. Aluminium sheet.

Materials suitable for jointing models

Glues and adhesives of all types. Pastes for cardboard and paper. Nails and pins. Screws. Nuts and bolts. Paper clips. Paper staples. Drawing pins. Needlework pins. Sellotape and masking tape. 'Blu-tack'. Soft iron wire. Fuse wire. Thread. Cotton. Metal parts may be soldered or welded.

Examples of models for technology

Twelve photographs showing examples of models for school technology are given on pages 141 to 143.

Models

The four photographs on this page show examples of the use of a variety of materials.

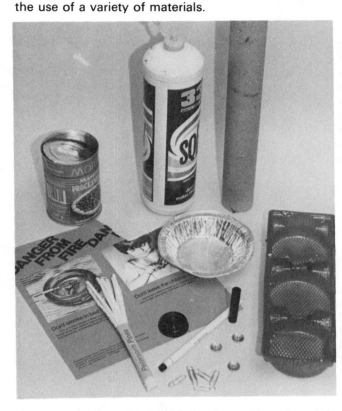

Any materials may be suitable for the making of models for technology. Shown here are—a food tin container; a washing-up squegee bottle made from plastics; a cardboard roll; a piece of cardboard; a tin lid and an old biro pen; an aluminium foil pastry can; some pipe cleaners; a used 'Penstik'; some paper clips, some drawing pins; a thin plastic pastry container.

A model hovercraft made from expanded polystyrene foam, paper and sticky tapes. Fitted to an electric motor to test working ability of the model.

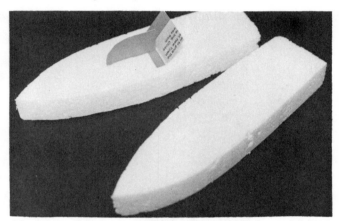

Model 'walking-on-water' floats made from expanded polystyrene foam and thin scrap cardboard. The card has been bent and glued to the underside of the models. Used to test whether the flap would work. See pages 26 to 29.

Cardboard from a food packet has been cut to shape and glued together to test whether the bubble blower tank fits into its allotted space. See pages 30 to 33.

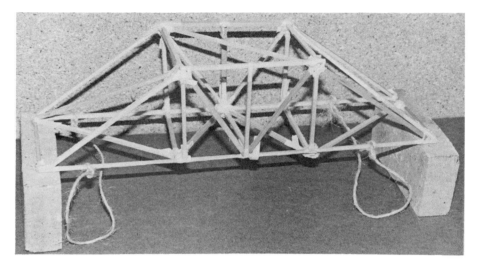

The use of strip wood (6 mm square) and masking tape glued joints to build model bridge structure.

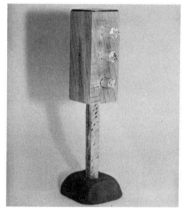

A model set of traffic lights made from wood mounted in a clay base with 'lights' made from a plastic pen barrel covered in silver paper. Used to test the layout positions of a model traffic control system.

A rough wooden case made to test whether batteries and circuit would fit into a bedroom light. See pages 14 to 17.

A model remote controlled gate made from any pieces of wood available, plus odd scraps of metal, a strip of Meccano and part of a Danum-Trent kit.

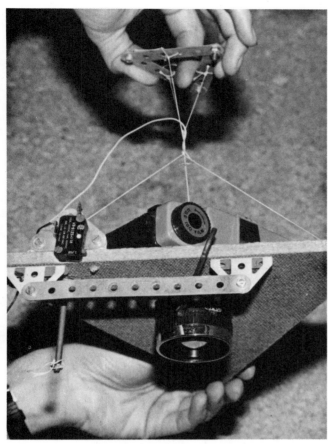

A gear box made from wood and acrylic sheet. A 'mock-up' to test a gearing system prior to constructing the finished design.

A 'mock-up' platform on which a camera is temporarily mounted in order to ascertain whether the 'cocking' and 'triggering' apparatus works before connecting a finished platform to a balloon.

A frame made from Meccano strips on which part of a device for automatically dispensing drinks is mounted. This is a 'mock-up' to test the device before making the finished design.

Testing a pneumatics circuit on a tracking system prior to mounting the circuit in a finished design.

Appraisal

The purpose of an appraisal of a design which has been made to meet a technological situation is to assess whether or not the design is successful. An appraisal should preferably be in writing, perhaps in note form. It could include drawings, graphs and statistics of performance among other details. When the written appraisal report has been completed, add it to the notes, drawings and other information held in the folder or folio containing all the material connected with the project.

A spoken appraisal, recorded on a cassette with the aid of a cassette player, could be considered appropriate as a form of recording an appraisal. The cassette could then be kept in the project folder when completed.

An appraisal report could include a photograph of the completed design, or if the design is one which should perform a variety of functions, several photographs may be considered necessary.

There should ideally be three parts to an appraisal report—tests; a report on the design; necessary modification.

Tests

Fully test the design under all the circumstances for which it has been made. Make notes of the results of the tests. If the design is one which performs or measures functions which can be measured such as— humidity, temperature, height, positions, timings and so on— graphs or other statistical drawings may be needed to clearly show the results of the tests. Then answer the following questions:
1. Do the tests show up faults in the design?
2. Can you give reasons for any faults which you have noted?

Report on the design

In order to give guidelines as to the contents of the report, the following questions could be asked and answered:
1. Does the completed design meet the requirements of the situation?
2. Does the design meet the demands of the design brief?
3. Have the applied tests shown that the design functions as expected?
4. Is the appearance of the design satisfactory? If not, is it necessary to re-design from the point of view of appearance? Perhaps the appearance does not matter.
5. Are there any safety requirements which the design fails to satisfy? Is the design safe to use?
6. Is the design economical in use? If it requires power for it to function, is it economical in the use of that power? Is it too large? Could it occupy a smaller space?
7. Is the design sufficiently strong for its purposes?
8. Are there any necessary maintenance problems which could arise when the design is put into constant use?

Modifications

Having tested and appraised the design, are modifications necessary or does the design need to be completely changed? Write down any modifications which may appear to be needed. Add drawings to your notes, so that when modifications have to be made, they can be clearly understood without having to test and report again on the design.

Economics

Your appraisal should include a note on the costs involved. What has the total cost of the project been? Do not forget to add such items as postage, cost of writing and drawing materials and other such incidentals incurred in the designing. Could the costs have been lowered without affecting the correct functioning of the design? Could cheaper, yet equally as efficient energy supplies have been used? Could different and cheaper materials have been employed without affecting the efficiency of the design? Look at the time taken in completing the design. Could you have made more efficient use of that time? Look at what you may have wasted in producing the design and, in so wasting, have increased the costs involved.

Conclusion

Your appraisal is a necessary part of the design process. It should be in a written form with possibly drawings and photographs. It can consist of three parts:
1. Tests
2. A report
3. Necessary modifications.

Your appraisal should consist of facts and not opinions. It is very easy to consider that a design you have just completed is excellent and to report accordingly. What is really required is to look at one's design very critically and be quick to explore the possibility that it is not as good as one would wish. If you cannot criticise your own design without being honest about its faults, ask other people to appraise the design and even ask them to write the appraisal report.

Exercises

Write appraisals of the following projects described in this book.
1. The CRDBTOIL—pages 10 and 11.
2. The bedroom light—pages 14 to 17.
3. The 100 km/h vehicle—pages 18 to 21.
4. The door opening device—pages 22 to 26.
5. The walking-on-water device—pages 26 to 29.

Examples of University projects

The following photographs were taken in the Mechanical Engineering Department of University College, London University.

Undergraduate project—Force and moments transducer. This device can measure three axes of force and three axes of moment. One can sit by a robot manipulator and measure the forces and moments applied at the end of the manipulator.

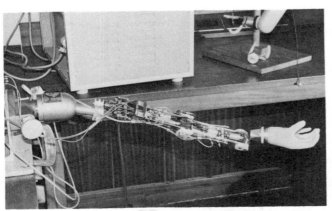

Staff and post-graduate project—Project to aid thalidomide children. Hydraulically operated arm. Hydraulic circuits in a back pack. Body powered arm on one side, hydraulically operated arm on other side.

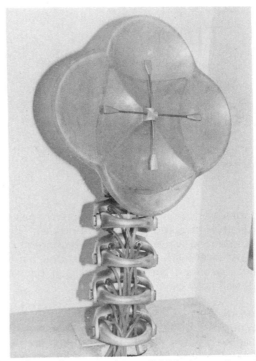

Staff and graduate project—SAM (sound activated mobile). Four microphones in parabolic reflectors. Vertical spine, hydraulically powered to bend, turn and move the reflectors towards the source of a sound.

Graduate project—A hydraulically operated robot arm.

Joint mechanism of the robot arm shown in the photograph above.

SECTION 4

Presentation

The pages concerned with solutions (pages 129 to 135) dealt with some of the methods by which the graphics needed when designing a technology project could be produced. Some suggestions as to how graphics based on these methods can best be presented, will now be considered.

Folders or design folios
All the graphics, notes, reference material, statistics etc., made or used when designing a project, are best kept together in a large envelope, a folder, in loose leaf covers or in a design folio. They should preferably be kept in an order of sequence and a simple number reference, or letter reference may be thought to be necessary on each sheet. A simple index to this sequence may also be desirable. The cover of the folder or folio should clearly state what it contains, and some graphical method of presenting the cover could be considered. Two examples of cover designs are shown. These are the covers for two of the projects already described in Section 1.

Two examples
The graphics relating to the three-minute timer are kept in simple home-made, A4 cardboard covers, the A3 drawing sheets being folded to fit. A photograph of the timer taken from its box is pasted on the cover. Titles in dry transfer print and a name and form number drawn with the aid of a lettering stencil are printed on grey paper and also pasted on the cover.

The material for the walking-on-water project is kept in an A3 manilla folder. A sheet of white card is pasted on its front. An outline drawing of one of the floats is drawn in indian ink with very thick outlines. This is again outlined by a blue colour wash. Titles and names are printed with dry tone transfer letters.

Dry tone transfer lettering
Some examples of dry tone transfer lettering are given, together with a photograph showing the letters being rubbed on to a drawing with the aid of a ball point pen. A pencil could be used for this purpose.

A huge variety of typefaces and size of print are available in dry transfer lettering sheets. Dry transfer lettering is also of value when wishing to print such details as ON:OFF positions; 1, 2, 3 etc. on the surfaces of electrical and electronics control boxes and similar parts of completed designs.

Stencil lettering
Some examples of lettering drawn with the aid of letter stencils are given, together with a photograph showing how a technical pen is used with a lettering stencil. There are a number of firms who manufacture lettering stencils. Each stencil contains all the letters of the alphabet in capitals and also in lower case, together with numerals and symbols such as brackets, commas, hyphens, question marks etc.

Paul Cartwright

Technology Project

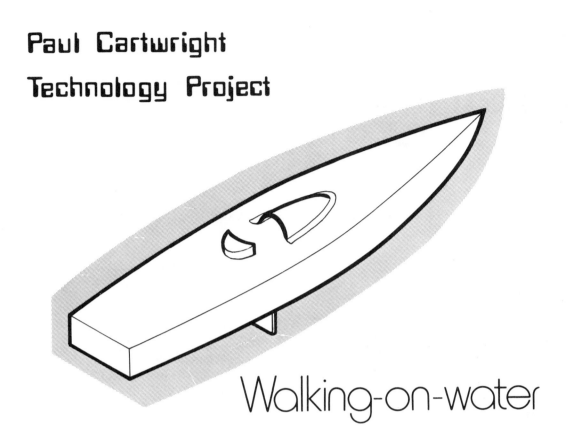

Walking-on-water

DSHRTUNO.
helvetica light
DSHRTUNO
baskerville old
DSRTU horatio light
OSHRTUNO   data 70

A B F H J L N P R T V Y   a b f h j l n p r t vy 12345
ABCDEFGHIJKLMNOPQRSTUVWXYZ 1234567890 ; ' &( )

A.YARWOOD   A.YARWOOD   A.YARWOOD
A.YARWOOD   A.YARWOOD   A.YARWOOD   A.Yarwood

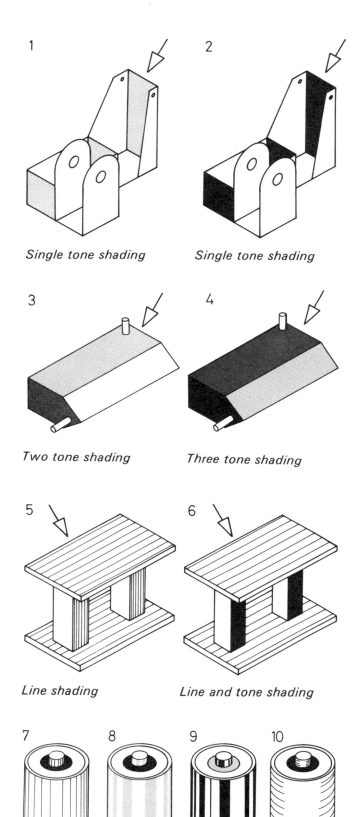

1

Single tone shading

2

Single tone shading

3

Two tone shading

4

Three tone shading

5

Line shading

6

Line and tone shading

7   **8**   **9**   **10**

Methods of shading cylindrical parts

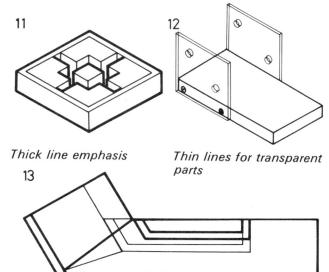

11

Thick line emphasis

12

Thin lines for transparent parts

13

Thicker lines show a modification

Presentation of graphics

The graphics needed when designing a technology project can be produced with the aid of instruments or can be drawn freehand. The line work, shading and colouring required may be drawn with:

Pencils—different hardnesses—2H, H, HB, B.
Coloured pencils—of the Caran d'Ache type.
Technical pens—eg 'Rotring' pens; line thicknesses of from 0.1 mm to 2 mm.
Ball tipped pens—black, blue, other colours.
Nylon tipped pens—black and other colours. 'Penstiks'.
Crayons—preferably *not* wax crayons.
Colour wash with brushes—water colours.
Dry tone transfer shading—eg 'Letratone'.

Examples of a variety of suitable techniques which can be adopted when designing a technology project are given here. It should be noted that these techniques can be employed when working with instruments or when drawing freehand.

Shading

Various methods of shading are shown in Drawings 1 to 10. These shadings can be applied with crayons, colour washes, dry-tone transfers or by pasting coloured paper where required. This type of shading can also be applied with B or HB pencils with varying pressures. Although a lighting position is assumed—as shown by arrows on some of the drawings—shading is added to drawings to assist in visualising the articles shown. Unshaded drawings may be preferable and the person producing the graphics must decide whether drawings are easier to visualise if shading is added.

Line emphasis

Drawings 11, 12 and 13 are illustrations in which graphical emphasis is produced by variation of line thicknesses.

148

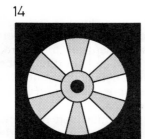

14

Black background

15

Shaded background

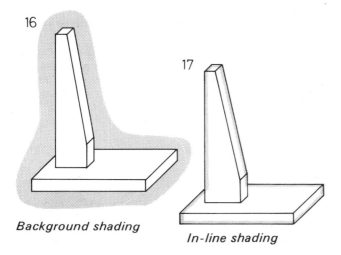

16

Background shading

17

In-line shading

entail some form of electrical, electronic or fluidic control. An emphasis can be given to a circuit by drawing on coloured paper and then pasting the paper on to a sheet to be included in a design folder or folio. Drawing 20 is an example of an electronics circuit on a coloured background.

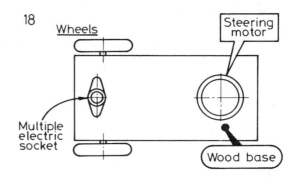

18

Labelling

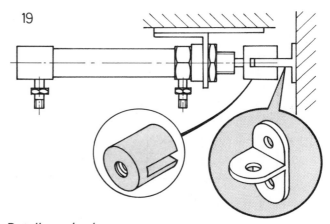

19

Detail emphasis

Background shading

Drawings 14 and 15 show black and grey background shading. Shadings such as these can be added to drawings to emphasise particular outlines. These shadings can be added with black ink, coloured inks, colour washes, crayons, shaded in with pencil, or by pasting cut-out drawings on to coloured paper. Dry tone transfers are also suitable for this purpose. Drawing 16 shows a common type of background shading.

In-line shading

Drawing 17 is an example of colour added just inside the outline of a drawing. Yellow crayon is particularly suitable for this purpose.

Labels

Drawing 18 gives a variety of methods of labelling parts of a drawing.

Detail extraction

Drawing 19 is an example of methods by which details in an orthographic view can be clearly explained by the addition of pictorial drawings showing the shape of the details.

Circuit diagrams

Circuit diagrams often play an important part in the design of technology projects, particularly those which

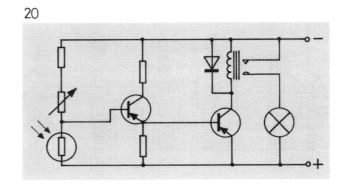

20

Circuit emphasis

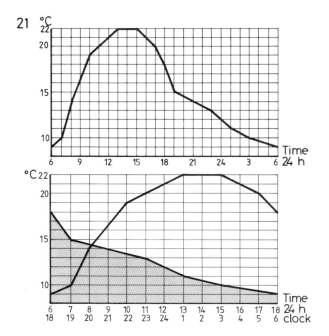

Line graphs

Graphs

The production of graphs to show results of experiments or as part of an assessment of the correct functioning of a project, can form an important part of the design of a project. A large variety of different forms of graphs can be drawn. Three types in common use are shown in Drawings 21, 22 and 23.

Drawing 21 shows two graphs which were produced in connection with the Aero-generator project described on pages 40 to 44, before the project was fitted into the greenhouse. Two methods of drawing the same graph—*line graphs*—the second with night and day temperatures compared on the same graph axes.

Drawing 22 is a *histogram*. This graph was part of a project designed to show how the application of a force can determine braking efficiency.

Drawing 23 shows a *graphical* method of showing comparisons on a yearly basis. This graph was part of a project designed to show how employment in an industrial concern varied with a variety of economic factors.

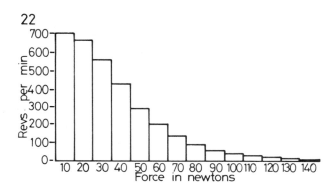

A histogram

Flow charts

Drawing 24 is a flow chart drawn with symbols taken from BS 4058 (specification for processing flow chart symbols, rules and conventions). It shows the design process as outlined throughout this book put into flow chart form. Flow charts may form a necessary part of a design for a technology project.

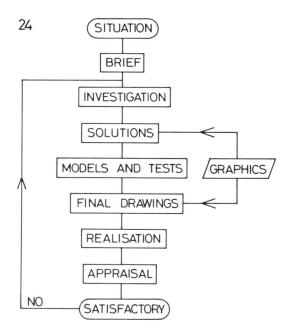

A flow chart

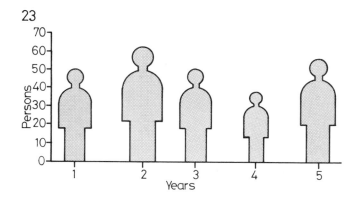

A graphical chart

The six photographs on this page show projects designed and made by 4th and 5th year pupils in secondary schools.

Part of a pneumatics system for a model fork lift truck.

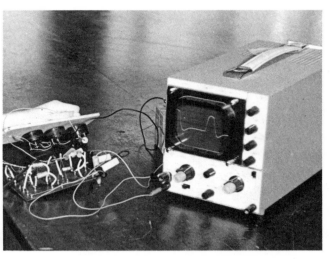

Testing the oscillation caused by a vibrator circuit.

Part of a device designed to control the food delivered to chickens.

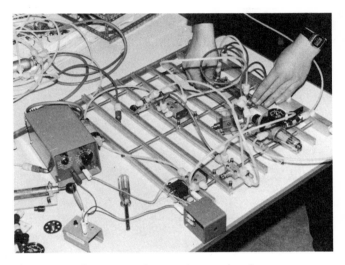

A pneumatics set up for testing a circuit.

A computer controlled vehicle—testing the program.

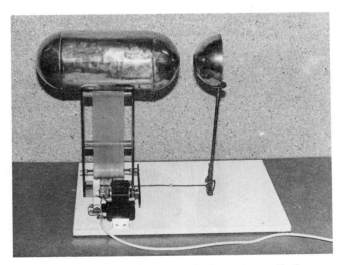

Part of a device for the control of smoke pollution.

Appendices
British Standard 308 drawing conventions

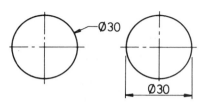

Dimensioning circles

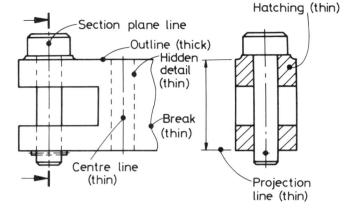

Lines used in drawings

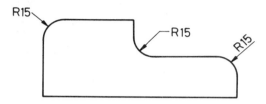

Dimensioning radii

First Angle Projection

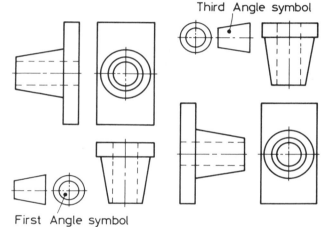

First Angle symbol

Third Angle symbol

Third Angle Projection

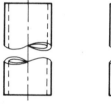

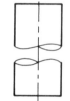

Break in a pipe   *Break in a rod*   *A spring*

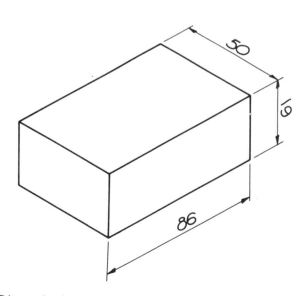

Dimensioning a pictorial drawing

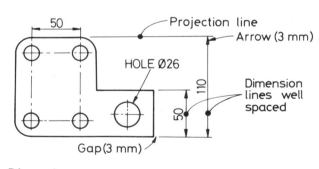

Dimensions

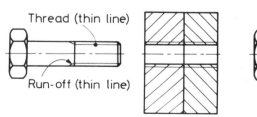

Screw threads

Electric and electronics symbols

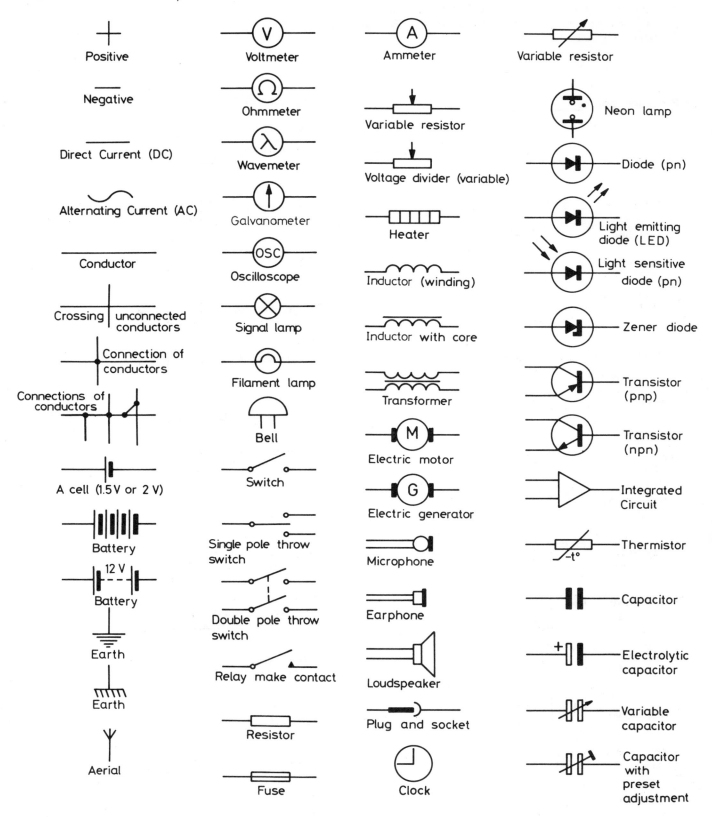

Positive

Negative

Direct Current (DC)

Alternating Current (AC)

Conductor

Crossing unconnected conductors

Connection of conductors

Connections of conductors

A cell (1.5 V or 2 V)

Battery

12 V Battery

Earth

Earth

Aerial

Voltmeter

Ohmmeter

Wavemeter

Galvanometer

Oscilloscope

Signal lamp

Filament lamp

Bell

Switch

Single pole throw switch

Double pole throw switch

Relay make contact

Resistor

Fuse

Ammeter

Variable resistor

Voltage divider (variable)

Heater

Inductor (winding)

Inductor with core

Transformer

Electric motor

Electric generator

Microphone

Earphone

Loudspeaker

Plug and socket

Clock

Variable resistor

Neon lamp

Diode (pn)

Light emitting diode (LED)

Light sensitive diode (pn)

Zener diode

Transistor (pnp)

Transistor (npn)

Integrated Circuit

Thermistor

Capacitor

Electrolytic capacitor

Variable capacitor

Capacitor with preset adjustment

153

Pneumatics symbols

A fuller explanation of the CETOP symbols used in pneumatics circuits will be found in the British Standard BS 2917 CETOP.

Pressure source

Exhaust

Working line

Control line

Exhaust line

Flexible pipe line

Line connection

Crossing lines

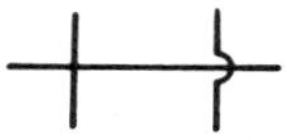

Pneumatic reservoir

2 way valve closed

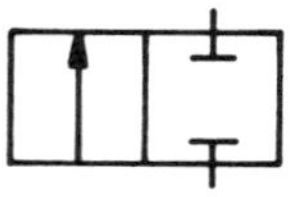

2 way valve open

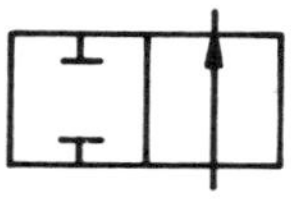

3 way valve closed

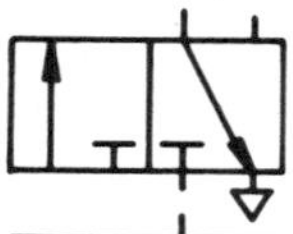

3 way valve open

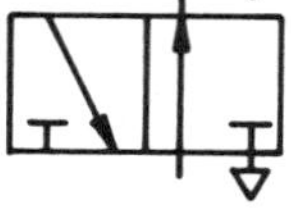

5 way valve

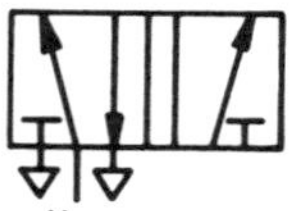

Letter labels on lines

Working lines   A,B,C,---

Power connection   P

Exhaust connections   R,S,T,---

Control lines   X,Y,Z,---

Valve position

Push button control

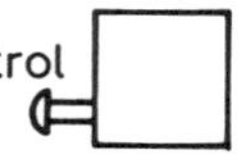

Lever control

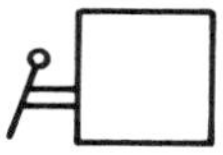

Pedal control

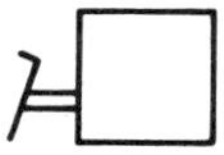

Plunger control

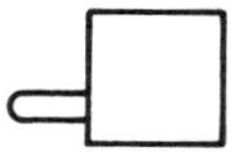

Spring control

Pneumatic control –pressure

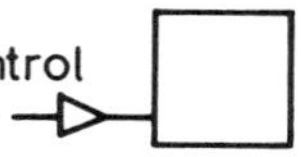

Pneumatic control –relief

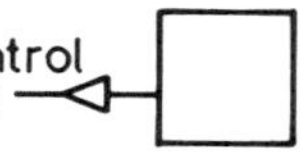

Flow meter

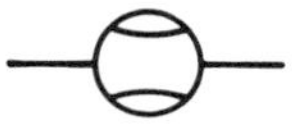

Single acting cylinder

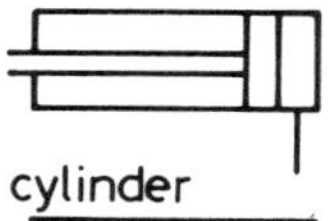

Single acting cylinder –spring return

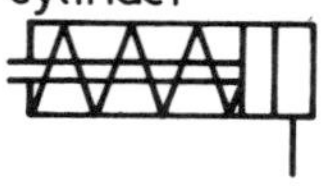

Double acting cylinder

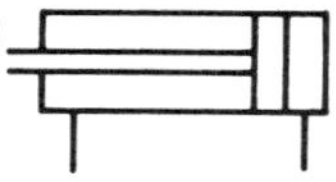

Restrictor

Variable restrictor

Shut-off valve

Exercises

Colour	Band 1	Band 2	Band 3	Band 4
Black	0	0	None	
Brown	1	1	0	1%
Red	2	2	00	2%
Orange	3	3	000	3%
Yellow	4	4	0 000	4%
Green	5	5	00 000	—
Blue	6	6	000 000	—
Violet	7	7	0 000 000	—
Grey	8	8	—	—
White	9	9	—	—
Gold	—	—	0.1	5%
Silver	—	—	0.01	10%
None	—	—	—	20%

1. A carbon resistor is shown. The colour code for such resistors is given in the table.

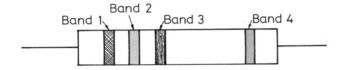

State the values and tolerances of the following resistors:

Band 1	Band 2	Band 3	Band 4
Red	Green	Orange	None
Brown	Black	Green	Silver

State the colour of the bands on the following resistors:
3.9 Ω, 10% tolerance; 100 Ω, 5% tolerance; 33 kΩ, 2% tolerance; 8.2 kΩ, 5% tolerance.

2. The bulbs specification for a car is:

2 sealed beam units headlight/dipped	60/45 W
2 side light/direction indicators	6/21 W
2 rear direction indicators	21 W
1 rear number plate	6 W
2 rear/stop lights	6/21 W
1 interior light	3 W
5 instrument panel lights	3 W

The battery is a 12 V, 38 ampere hour type

Questions (i) to (v) are based on this bulbs specification.

(i) Calculate the current drawn when testing right turn indicator bulbs with the engine not running.

(ii) Calculate the current drawn testing side and rear lights with rear number plate and all panel lights on with engine not running.

(iii) If the interior light is left on for five whole days, how many ampere hours would be consumed?

(iv) Calculate the resistance of the 6 W number plate bulb.

(v) If you measured the resistance of the 6 W number plate bulb with an ohmmeter would you obtain the same reading as the calculated resistance? Give a reason for your answer.

3. (i) A resistor of 750 Ω is required, but you have only 1 kΩ resistors. Draw a sketch showing how a resistor of 750 Ω could be made up from the 1 kΩ resistors.

(ii) What is the maximum steady current that should be passed through a 100 Ω, 1 W resistor?

4. A bicycle dynamo is mounted on a board and connected to a lamp and a centre reading galvanometer. The dynamo is turned by the wheel W.

(i) Explain all the things you would observe at the lamp and galvanometer when the wheel is being turned slowly.

(ii) When the rotation of the wheel is increased so that the dynamo is turning at its normal speed, the lamp is very bright but the galvanometer does not appear to register. Explain these observations.

(iii) Why is the circumference of the wheel larger than that of the pulley on the dynamo?

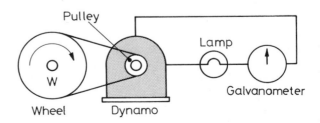

5. The diagram shows a circuit often used for plotting transistor characteristics in common emitter connection. Study the circuit and answer the questions.

(i) What do the letters e, b and c stand for?

(ii) Which type of transistor is being investigated?

(iii) The two ammeters to be connected at X and Y are a microammeter and a milliammeter. Which would you insert at X? Briefly explain why.

(iv) What is the purpose of R_2?

(v) Why is it necessary to connect another resistor R_1 in series with R_2?

(vi) Give a reason why the voltmeter at Z should have a high resistance.

(vii) What does R_3 control? What is the name given to a resistor used in this manner?

(viii) The transistor is supported in a clip or inserted in a cylindrical hole in a block of copper. Explain why it is desirable to do this.

(ix) Sketch the shape of the output characteristics that would be obtained with base current constant.

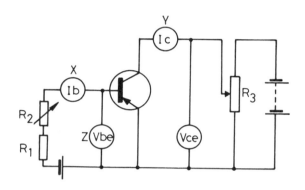

6. A bulb is marked 12 V, 36 W. What is the resistance of the bulb filament when it is connected to a 12 V supply?

7. A rheostat is marked 10 A, 200 W. This means that it can safely give out heat at the rate of 200 W when a maximum current of 10 A is flowing through it.

(i) What is the largest potential difference that can be put across the terminals of the rheostat?

(ii) When a current of 10 A is flowing how much heat energy will be given out in one minute?

8. An object has a mass of 5 kilograms and is resting on a horizontal surface. It is found by experiment that a force of 8 newtons, acting in a horizontal direction, is needed to just start the object moving.

(i) Why will an applied force of 3 newtons not move the object?

(ii) Why isn't a force of 50 newtons needed to move the object?

(iii) Give an approximate size of the force that will just keep the object moving with constant velocity.

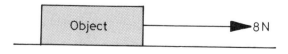

9. A rod is acted upon by five vertical forces as shown to form a 'mobile'. Ignoring the mass of the rod, find the exact position of suspension point A for the system to be in balance.

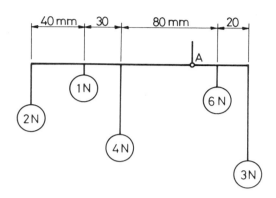

10. The plane STUVW is freely pivoted centrally along edge ST at B. Forces of 5 N, 4 N and 6 N are applied vertically downwards to corners U, V and W. The vertical rod AB is fixed at A. At what angle to the horizontal will ST tilt when the system is in balance?

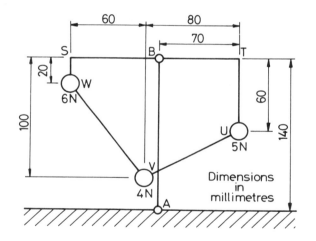

11. Copy the four diagrams and draw what happens to the ray of light in each case.

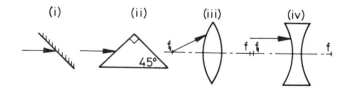

12. The following diagrams represent the action of different optical devices upon the two rays of light A and B. Each optical device is 'hidden' in the dotted outline.

From the drawings shown below decide which optical device best fits each box.

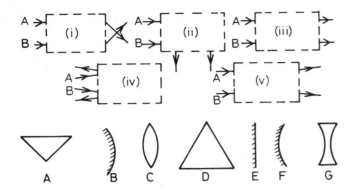

13. When driving on the Continent, British motorists use a plastics sheet to fit over their headlamps so that they dip to the right. With the aid of the diagram below explain how this arrangement works.

14. Explain why a piece of red plastics shaped as shown above makes a better reflector for the back of a vehicle than a piece of highly polished metal.

15. (a) Polymers are either 'thermoplastic' or 'thermosetting'. Explain the meanings of these two terms.
(b) Name three polymers which are 'thermoplastic' and three which are 'thermosetting'. State a use to which each of the polymers you have named can be put.
(c) Thermoplastics can be 'injected', 'extruded' or 'calendered' to produce various products. Show by diagrams what is meant by these three terms and briefly explain the processes involved.
(d) Make a list of the processes involved in correct order of sequence in the 'laying-up' of a fibre-glass moulding.

16. A conventional type of rainfall recorder is shown. Every 24 hours the level of the water in the graduated glass cylinder must be measured and recorded. The cylinder is then emptied ready to receive the next 24 hours rainfall.
(a) Name other devices usually associated with the recording of details of weather.
(b) A conversion to the system shown is required in which the daily recordings of rainfall can be automatically recorded inside a building making use of an electrical meter. Design one method by which this can be achieved.

(c) Also design a method by which the cylinder can be emptied from within the building by remote control.

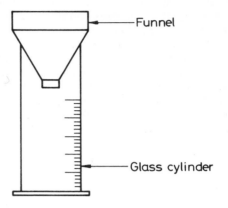

17. (a) Sketch a graph showing the results of a tensile test for steel. Label the x and y axes, including units of measurement.
(b) Sketch a bar graph showing the results of impact tests on four different materials. Name the materials, label the x and y axes and include units of measurement.
(c) To enable results to be read clearly, materials testing machines usually require some means by which the physical results of tests can be enlarged or amplified. Suggest two methods by which the results of the hardness tester shown can be so amplified.
(d) Sketch the better of the two methods you have stated and explain how it works. Indicate by how much you expect it to amplify.

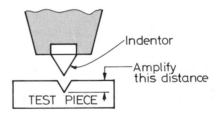

18. (a) Why is geothermal energy often considered to be an income energy source even though its life is limited?
(b) Many advertisements claim that 'solar energy' is 'free'. Explain briefly why this statement is misleading.
(c) One of the main arguments against total energy schemes such as the Severn Barrage is the damage they may cause to the environment and animal life. Give two examples of this, and briefly explain what damage can be caused.　　　(SUJB)

19. The mist propagator shown is to be fitted in a greenhouse. Water is sprayed over seedlings automatically when the sensor is dry and ceases when the sensor is too wet.
(a) The following are essential to the successful functioning of the mist propagator—valve control; relay; sensor; amplifier. Place the four in a correct order for the propagator to function.
(b) Draw an electronics (or other) circuit which could be used to control the propagator. You could use a simple one or two transistor amplifier in the circuit. Do not put any values on your components, but use the correct symbols appropriate to your circuit. Explain briefly how the circuit functions.

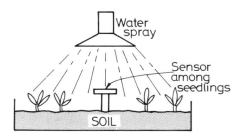

20. The drawing shows a section through the top of a large central heating boiler that is to be installed into a new school.

In order that the boiler may function without exploding, a safety valve must be fitted that will allow a build-up of a maximum pressure of 1 Nmm^{-2} before releasing excess pressure.

Taking atmospheric pressure as 0.1 Nmm^{-2} design and justify by calculation a safety valve based on a lever system which will meet the above requirements.

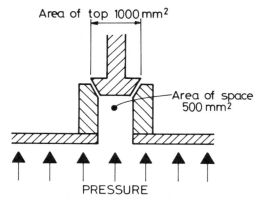

21. The adhesion of surface coatings, such as electroplate, paints and varnishes is an important industrial technology. One test for the effectiveness of such adhesion is the Sellotape test. This involves a strip of Sellotape being pressed on to the surface coating and then measuring the force required to strip off the surface. The drawing shows this test in a diagrammatic form.
  (i) If the pulling force is derived by means of weights and a pulley, calculate the size of the pulley load to give a pull on the specimen of 50 000 newtons.

(ii) What is the mechanical efficiency of this machine?
(iii) What is the velocity ratio of the machine?

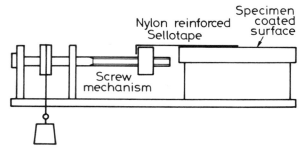

22. The drawing shows an incomplete mechanical arrangement driven by an electric motor.
(a) Sketch and name a suitable coupling that could be used at A.
(b) Make calculations to deduce the gears required at B if the shaft C is required to rotate at 7500 rev/min.
(c) Sketch and name a device which could be used at C to transmit the motion through 90° without changing the speed of rotation.
(d) Make a clear sketch of an arrangement that could be used to transmit power from the shaft to the pulley in a fixed position on the shaft.
(e) Name the bearing shown at E and comment critically on the seating design.
(f) The double-pulley arrangement at F and G is to be belt-driven from the pulley at D. The diameter of the pulleys are: D, 100 mm; F, 150 mm; G, 250 mm. Calculate the speed of rotation of the shaft H in revolutions per minute for each of the two possible positions of the belt.   (Oxford)

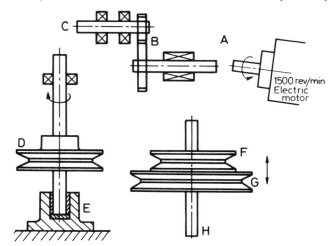

23. A refrigerator has a motor designed to operate for 48 000 hours without needing replacement. To replace the motor costs about half the cost of a new refrigerator.
(a) Explain how these facts illustrate the idea of 'inbuilt obsolescence'.
(b) Give *two* advantages that inbuilt obsolescence can have.   (SUJB)

Index

Acknowledgements

We wish to express our appreciation of the assistance given by the representatives of the following industrial undertakings and organisations for allowing copyright photographs to be reproduced in this book: J. C. Bamford (Excavators) Limited of Rocester, Staffordshire for allowing us to take photographs in their factory for reproduction in this book; Freeman Fox and Partners for supplying photographs of the Humber Bridge and of the Parks Radio Telescope; The British Railways Board for supplying photographs of the Advanced Passenger Train; The British Aircraft Corporation for supplying photographs of Concorde; Ferranti Electronics Limited for the photograph of the circuits of their silicon chip F100–L; IBM (United Kingdom) Limited who supplied photographs of computer applications; International Computers Limited who also supplied photographs of computer applications; Central Laboratories, SEGAS, for supplying a photograph of their experimental 'Mains Buggy'. Our thanks are also due to the lecturers, teachers and pupils in colleges and schools who gave so freely of their time and knowledge in order to assist us in compiling this book. Among others, we wish to record our gratitude to: Mr. R. Preston, a lecturer at the North Staffordshire Polytechnic and a technician at the college, Mr. Derek Cheers, both of whom devoted time to ensuring that photographs could be taken of testing devices at their college; Mr. N. A. Brittain and pupils at the Walton High School, Stafford, for allowing details of their aero-generator project to be published in this book; the pupils at the Thomas Alleyne's High School, Uttoxeter, who gave their time so freely to assist in the making of projects. In particular our thanks are due to Jon Kirk for taking photographs from which Jean Yarwood produced the drawings of craft processes for pages 114 to 121.

The undermentioned examination boards have kindly allowed questions from examination papers set by them in recent years to be copied for reproduction in this book:
The Associated Examining Board;
Oxford Delegacy of Local Examinations;
The Southern Universities Joint Board.

A. Yarwood.
A. H. Orme.